In the Matter of
J. Robert Oppenheimer

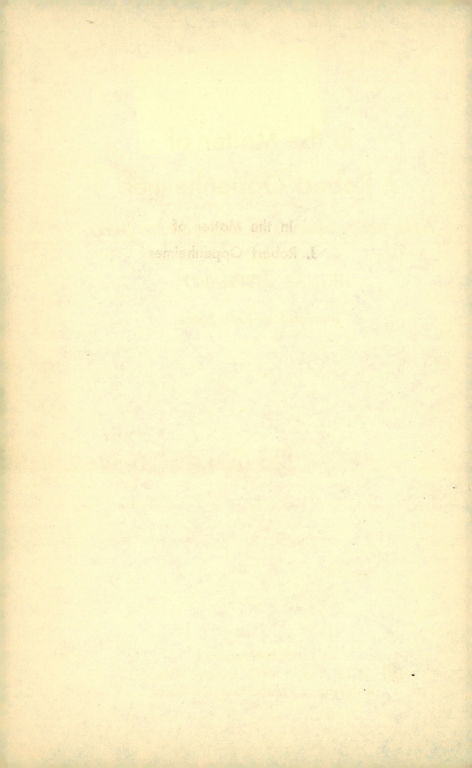

In the Matter of
J. Robert Oppenheimer

A play freely adapted on the basis of the documents
by
HEINAR KIPPHARDT

Translated by Ruth Speirs

A *Spotlight Dramabook*
HILL AND WANG · NEW YORK
A *division of Farrar, Straus and Giroux*

Originally published in German as
In der Sache J. Robert Oppenheimer.
© 1964 Suhrkamp Verlag, Frankfurt am Main. All rights reserved.
This translation © 1967, 1968 by John Roberts
All rights reserved
Standard Book Number (clothbound edition): 8090–5770–0
Standard Book Number (paperback edition): 8090–1215–4
Library of Congress catalog card number: 68–18845

First Dramabook edition December 1968
Second printing February 1969

Manufactured in the United States of America
by The Colonial Press Inc., Clinton, Massachusetts

7   8   9   10

The Play in Relation to the
Documentary Data

In the Matter of J. Robert Oppenheimer is a play for the theatre, not an assemblage of documentary material. Even so, the author adheres strictly to the facts which emerge from the documents and reports concerning this investigation.

His chief source is the records—three thousand typewritten pages—of the proceedings instituted against J. Robert Oppenheimer; these records were published by the United States Atomic Energy Commission in May 1954.

It is the author's intention to present a shortened version of those proceedings, a version which lends itself to being staged and which does not distort the truth. As the author's business is the stage, and not the writing of history, he endeavors to follow Hegel's advice and lay bare "the core and significance" of a historical event by freeing it from the "adventitious contingencies and irrelevant accessories of the event," to "strip away the circumstances and aspects that are of merely secondary importance, and to replace them with such that allow the essence of the matter to appear in all its clarity." (Hegel, *Aesthetik* Part III, Chapter 3, p. 897; Berlin, 1955.)

Even so, after mature consideration, the author deliberately confined himself to drawing only upon historical data for all the facts presented in this play. The author exercised his freedom only in the selection, the arrangement, formulation, and condensation of the material. Some filling-in and intensification was necessary in order to achieve a more tightly knit as well as more comprehensive documentation and, as such, more appropriate for the stage. In this respect, the author was guided by the principle: as little as possible, and as much as is indispensable. When the truth seemed jeopardized by an effect, he sacrificed the effect.

5

A few of the liberties taken by the author: The original hearing took over a month, and forty witnesses were called. The author confined himself to six witnesses. It was impossible to achieve the required concentration with a verbatim reproduction of statement and counterstatement, nor did this seem desirable to the author in the interests of the unity of the play. Therefore, he endeavored to subordinate word-for-word recapitulation to accuracy of meaning.

As a result of the author confining himself to six witnesses, it occasionally happens in the play that complementary depositions are merged into the evidence put forward by one single witness. For instance, the simile of the bank robbery was in actual fact introduced by Robb, not by Morgan; Robb, in this connection, was questioning the witness McCloy—not Lansdale, as in the play.

The author has introduced monologues by the dramatis personae between scenes of the play; there were no such monologues at the actual proceedings. He has tried to evolve these monologues from the attitudes adopted by these persons in the course of the proceedings or on other occasions.

Some of the thoughts expressed by Teller as presented on the stage were, though not verbatim, culled by the author from speeches and articles by Teller. Oppenheimer had three lawyers; in the play, he has two. Herbert S. Marks, who acts as his counsel throughout the play, was engaged to advise Oppenheimer only in the course of the proceedings. The speech for the defense was made by Garrison—not by Marks, as in the play.

The final decision of the Personnel Security Board was not read out at the end of the proceedings, as it is in the play, but was subsequently made known by letter. Dr. Oppenheimer did not make any closing statement. He was given the opportunity to do so at the conclusion of the hearings but he used it solely to make a point of a technical nature.

HEINAR KIPPHARDT

In the Matter of
J. Robert Oppenheimer

CHARACTERS

J. ROBERT OPPENHEIMER, *physicist*

Personnel Security Board

GORDON GRAY, *Chairman*
WARD V. EVANS, *member*
THOMAS A. MORGAN, *member*

Counsel

ROGER ROBB, *counsel for the Atomic Energy Commission*
C. A. ROLANDER, *associate of Robb, security expert*
LLOYD K. GARRISON, *counsel for Oppenheimer*
HERBERT S. MARKS, *counsel for Oppenheimer*

Witnesses

MAJOR NICHOLAS RADZI, *Security Officer*
JOHN LANSDALE, *lawyer, formerly Security Officer*
EDWARD TELLER, *physicist*
HANS BETHE, *physicist*
DAVID TRESSEL GRIGGS, *Chief Scientist of the Air Force, geophysicist*
ISADOR ISAAC RABI, *physicist*

PART ONE

The stage is open. Visible spotlights. White hangings separate the stage from the auditorium, sufficiently high for the following documentaries to be projected on them:

Scientists in battledress, looking like military personnel, are doing the count-down for test explosions—4–3–2–1–0 (in English, Russian, and French).

Cloud formations caused by atomic explosions unfold in great beauty, watched by scientists through dark filters.

On the wall of a house, radiation shadows of a few victims of the atomic explosion on Hiroshima.

The hangings open.

SCENE 1

Room 2022

A small ugly office; walls of white-washed wooden boards. The room has been temporarily furnished for the purpose of the investigation. On a raised platform, back center, a table and three black leather armchairs for the members of the Board. Behind, on the wall, the Stars and Stripes. In front of the platform, floor level, the stenographers are seated with their equipment. On the right, ROBB and ROLANDER, counsel for the Atomic Energy Commission, are busying themselves with stacks of documents. Opposite ROBB and ROLANDER, on a raised platform, tables and chairs for OPPENHEIMER's counsel. In front of the platform, floor level, a small old leather sofa.

J. ROBERT OPPENHEIMER enters Room 2022 by a side door on the right. He is accompanied by his two lawyers. An official leads him diagonally across the room to the leather sofa. His

9

lawyers spread out their materials. He puts down his smoking
paraphernalia and steps forward to the footlights.

OPPENHEIMER. On the twelfth of April 1954, a few minutes to
ten, J. Robert Oppenheimer, Professor of Physics at Princeton, formerly Director of the Atomic Weapons Laboratories
at Los Alamos, and, later, Adviser to the Government on
atomic matters, entered Room 2022 in Building T3 of the
Atomic Energy Commission in Washington, to answer
questions put to him by a Personnel Security Board, concerning his views, his associations, his actions, suspected of
disloyalty. The evening before this investigation, Senator
McCarthy said in a television interview:

A huge picture of Joseph McCarthy is projected on the white
screens at the back. OPPENHEIMER *goes to the leather sofa*
and fills his pipe. A voice shaking with agitation issues from
the loudspeakers.

McCARTHY'S VOICE. If there are no Communists in our government, why do we delay the hydrogen bomb by eighteen
months while our defense services report day after day that
the Russians are feverishly stepping up on the H-bomb?
Now they've got it! Now our monopoly is gone! When I
tell America tonight that our nation may perish, it will
perish because of that delay of eighteen months. And, I ask
you, who is to blame? Were they loyal Americans or were
they traitors, those who deliberately misled our government,
who got themselves celebrated as atomic heroes, and whose
crimes must at last be investigated.

The members of the Board enter by a small door, back center.
Those present rise for a moment. Then everybody sits down.

GRAY. This Board has been appointed by the United States
Atomic Energy Commission to investigate Dr. J. Robert
Oppenheimer's continued eligibility for clearance. It is com-
posed of the following members: Thomas A. Morgan,
Ward V. Evans, and myself, Gordon Gray, Chairman.
Counsel for the Atomic Energy Commission are Roger Robb
and C. A. Rolander. Dr. Oppenheimer is represented by
Lloyd K. Garrison and Herbert S. Marks. Dr. Oppenheimer

is present as a witness in his own case. This inquiry is not a trial. It shall be regarded as strictly confidential.

MARKS. May I ask, Mr. Chairman, whether any of you saw the interview with Senator McCarthy last night?

GRAY. I did not see it. Did you, Mr. Morgan?

MORGAN [*looking up from his documents for a moment*]. McCarthy? No.

EVANS. I heard it on the radio. I was greatly surprised. I immediately thought of Oppenheimer.

MARKS. Did you hear the interview, Mr. Robb?

ROBB. No. Senator McCarthy would have to be clairvoyant if he alluded to our proceeding.

MARKS. He was interviewed by Fulton Lewis, Jr. I believe you represented that gentleman at various trials, Mr. Robb.

GRAY. Did you take his remarks as referring to yourself, Dr. Oppenheimer?

OPPENHEIMER. Five or six people called me up. Einstein said: "If I had the choice again I'd rather be a plumber or a pedlar, if only to enjoy some small measure of independence."

MARKS. I mention the interview because it makes me wonder if our proceedings can be kept private, Mr. Chairman.

GRAY. We shall do our best. . . . It is my duty, Dr. Oppenheimer, to ask whether you are satisfied with the composition of the Board.

OPPENHEIMER. Yes. With one general reservation.

GRAY. What is your reservation?

OPPENHEIMER. The Board will examine the complex duties of a physicist in our times; therefore, I would have preferred the members to be scientists. Only Professor Evans is engaged in science, I believe.

EVANS. But I don't know anything about nuclear physics, either. Fortunately. You probably know that we had no choice in this matter here. We were appointed. I wouldn't have chosen it myself.

OPPENHEIMER. Neither would I, I guess.

MARKS. The profession of the members should perhaps be shown in the record.

GRAY. Very well, Mr. Marks. Ward V. Evans . . .

EVANS. Professor of Chemistry, Chicago.

GRAY. Thomas A. Morgan . . .

MORGAN. Chairman of the Board and President of the Sperry Gyroscope Company, atomic equipment. One of the sharks of Big Business. [*He laughs.*]

GRAY. Gordon Gray, newspaper editor, radio stations; former Secretary of the Army, Department of Defense.

MORGAN. Information concerning our income is not required?

MARKS. You wouldn't want to disclose yours, Mr. Morgan.

Slight laughter.

GRAY. I would like to ask Dr. Oppenheimer whether he wishes to testify under oath.

OPPENHEIMER. Certainly.

GRAY. You are not obliged to do so.

OPPENHEIMER. I know. [*He rises to his feet.*]

GRAY. Julius Robert Oppenheimer, do you swear that the testimony you are to give the Board shall be the truth, the whole truth, and nothing but the truth, so help you God?

OPPENHEIMER. I do.

GRAY. The proceeding may now commence. May I ask you to take the stand . . . Mr. Robb.

OPPENHEIMER *walks across to a swivel chair which faces the members of the Board. He sits down and lights his pipe.*

ROBB. You have been called the Father of the Atom Bomb, Doctor?

OPPENHEIMER. In magazines. Yes.

ROBB. You would not call yourself that?

OPPENHEIMER. It isn't a very pretty child—and it has about a hundred fathers, if we consider the basic research. In several countries.

ROBB. But the baby was ultimately born in Los Alamos, in

the laboratories which you yourself had set up, and of which you were the Director from 1943 to 1945.

OPPENHEIMER. We produced that patent toy, yes.

ROBB. So you are not denying it, Doctor? [OPPENHEIMER *laughs.*] You produced it in a fantastically short time, you tested it, and then you dropped it on Japan, did you not?

OPPENHEIMER. No.

ROBB. You did not?

OPPENHEIMER. The dropping of the atom bomb on Hiroshima was a political decision—it wasn't mine.

ROBB. But you supported the dropping of the atom bomb on Japan. Or didn't you?

OPPENHEIMER. What do you mean by "supported"?

ROBB. You helped to select the targets, did you not?

OPPENHEIMER. I was doing my job. We were given a list of possible targets . . .

ROBB. Would you name them?

OPPENHEIMER. Hiroshima, Kokura, Nigata, Kyoto . . . [*Partial views of these cities are projected on the white screens at the back.*] . . . and we, as experts, were asked which targets would be most suitable for the dropping of the atomic bomb, according to the experience we had gathered from tests.

ROBB. Whom do you mean by "we," Doctor?

OPPENHEIMER. An advisory council of nuclear physicists, appointed for this purpose by the Secretary of War.

ROBB. Who was on that council?

OPPENHEIMER. Fermi, Lawrence, Arthur H. Compton, and myself.

Photographs of these scientists are projected on the screens.

ROBB. And you had to select the targets?

OPPENHEIMER. No. We supplied the scientific data as to the suitability of the targets.

ROBB. What kind of target did you consider to be of the desired suitability?

OPPENHEIMER. According to our calculations, the area had to be at least two miles in diameter, densely built up, preferably with wooden buildings—because of the blast, and the subsequent wave of fire. Also, the selected targets had to be of a high military and strategic value, and unscathed by previous bombardments.

ROBB. Why, Doctor?

OPPENHEIMER. To enable us to measure exactly the effect of a single atomic bomb.

EVANS. These military considerations, I mean, after all, they were the business of the physicists, weren't they, at that time?

OPPENHEIMER. Yes. Because we were the only people who had the necessary experience.

EVANS. I see. I'm rather out of my depth here. How did you feel?

OPPENHEIMER. I asked myself that question, later. I don't know. . . . I was very relieved when the Secretary of War followed our suggestions and crossed the famous temple city, Kyoto, off the list. It was the largest and most vulnerable target.

ROBB. But you did not oppose the dropping of the atom bomb on Hiroshima?

OPPENHEIMER. We set forth arguments against . . .

ROBB. I am asking you, Doctor, whether *you* opposed it.

OPPENHEIMER. I set forth arguments against dropping it.

ROBB. Against dropping the atom bomb?

OPPENHEIMER. Yes, that's right. But I did not press the point. Not specifically.

ROBB. You mean to say that having worked day and night for three or four years to produce the atomic bomb, you then argued it should not be used?

OPPENHEIMER. No. When I was asked by the Secretary of War I set forth the arguments both for and against. I expressed my uneasiness.

ROBB. Did you not also determine the height, Doctor, at

which the atomic bomb was to explode in order to produce the maximum effect?

OPPENHEIMER. We, as experts, were doing a job we were asked to do. But this does not mean that we thereby decided that the bomb should in fact be dropped.

ROBB. You knew of course, did you not, that the dropping of the atomic bomb on the target you had selected would kill thousands of civilians?

OPPENHEIMER. Not as many people as we thought, as things turned out.

ROBB. How many were killed?

OPPENHEIMER. Seventy thousand.

ROBB. Did you have moral scruples about that?

OPPENHEIMER. Terrible ones.

ROBB. You had terrible moral scruples?

OPPENHEIMER. I don't know anyone who would *not* have had terrible moral scruples after the dropping of the bomb.

ROBB. Isn't that a trifle schizophrenic?

OPPENHEIMER. What is? To have moral scruples?

ROBB. To produce the thing, to pick the targets, to determine the height at which the explosion has the maximum effect —and then to be overcome by moral scruples at the consequences. Isn't that a trifle schizophrenic, Doctor?

OPPENHEIMER. Yes. . . . It is the kind of schizophrenia we physicists have been living with for several years now.

ROBB. Would you elucidate that?

OPPENHEIMER. The great discoveries of modern science have been put to horrible use. Nuclear energy is not the atomic bomb.

ROBB. You mean it could be exploited industrially, and so forth?

OPPENHEIMER. It could produce abundance, for the first time. It's a matter of cheap energy.

ROBB. Are you thinking of a Golden Age, a Land of Cockaigne, that sort of thing?

OPPENHEIMER. Yes, plenty for all. It is our misfortune that people rather think of the reverse kind of uses.

ROBB. Whom do you mean by "people," Doctor?

OPPENHEIMER. Governments. The world is not ready for the new discoveries. It is out of joint.

ROBB. And you have come along "to set it right," as Hamlet says?

OPPENHEIMER. I can do no such thing. The world itself must do that.

MORGAN. Dr. Oppenheimer, do you mean to tell an old pragmatist like me that you made the atomic bomb in order to create some Land of Cockaigne? Did you not make it in order to use it and win the war with it?

OPPENHEIMER. We made it in order to prevent it being used. Originally, at any rate.

MORGAN. You spent two billion dollars of the taxpayers' money on the bomb in order to prevent it being used?

OPPENHEIMER. To prevent it being used by Hitler. In the end it turned out that there wasn't any German atomic bomb project. . . . But then we used it all the same.

ROLANDER. I beg your pardon, sir, but were you really not asked—at a certain stage in the development of the bomb —were you not asked whether it should be used against Japan?

OPPENHEIMER. We weren't asked *whether* it should be used, but only *how* it should be used in order to produce the maximum effect.

ROLANDER. Is that entirely correct, sir?

OPPENHEIMER. What do you mean?

ROLANDER. Did not the Secretary of War show you the so-called Franck Report, the memorandum by the physicists Szilard, Franck, and others? It strongly opposed the dropping of the bomb on Japan and recommended an internationally public demonstration of the bomb over a desert.

OPPENHEIMER. We were given it to read. That's right. Not officially, I believe.

ROBB. What did you say to that, Doctor?

OPPENHEIMER. That we were in no position to decide this question, that opinion was divided among us. We set forth our arguments—for, and against.

ROBB. Were you against?

OPPENHEIMER. Lawrence was against. I was undecided, I'm afraid. I think we said that the exploding of one of these things as a firecracker over a desert wasn't likely to be very impressive—and, probably, that the overriding considera- tion should be the saving of lives, by bringing the war to an end as soon as possible.

ROBB. Did this not mean, in effect, Doctor, that you were *against* a demonstration of the weapon—and *for* it being dropped without warning?

OPPENHEIMER. It most certainly did not mean that. No. We were physicists, not the military, not politicians. That was the time of very heavy fighting on Okinawa. It was a hor- rible decision.

ROBB. Did you write the official report on the effect of the bomb on Hiroshima?

OPPENHEIMER. According to the data supplied by Alvarez, yes; he flew in with the others, to measure the effect.

EVANS. Alvarez the physicist?

OPPENHEIMER. Yes. With new measuring instruments.

ROBB. Did you not state there that the dropping of the bomb had been a good thing, and very successful?

OPPENHEIMER. It was technically successful, yes.

ROBB. Oh, technically. . . . You are very modest, Doctor.

OPPENHEIMER. No, I am not.

ROBB. You are not?

OPPENHEIMER. We scientists have been on the brink of pre- sumptuousness in these years. We have known sin.

ROBB. Good, Doctor. We shall speak of those sins.

OPPENHEIMER. I guess we don't mean the same thing.

ROBB. That is something we are going to find out, Doctor. . . .

The reason I am digging up this old Hiroshima business is this: I want to find out why, at that time, you devoted yourself with such single-mindedness to your tasks, with a hundred-per-cent loyalty, I would say—and why, later, in the matter of the hydrogen bomb, you adopted an entirely different attitude.

OPPENHEIMER. It doesn't bear comparison, I think.

ROBB. It doesn't?

OPPENHEIMER. No.

ROBB. Would you have supported the dropping of a hydrogen bomb on Hiroshima, Doctor?

OPPENHEIMER. It would have made no sense at all.

ROBB. Why not?

OPPENHEIMER. The target was too small. . . . We were told that the atom bomb was the only means of bringing the war to an end quickly and successfully.

ROBB. You don't have to defend yourself, Doctor. Not on that count, anyway.

OPPENHEIMER. I know.

ROBB. Did the allegations contained in the letter of the Atomic Energy Commission surprise you?

OPPENHEIMER. They depressed me.

ROBB. What exactly depressed you, Doctor?

OPPENHEIMER. That twelve years' scientific work in the service of the United States should end in such allegations. . . . Twenty-three points in that letter deal with my associations with Communists or Communist sympathizers, associations going back more than twelve years. The letter contains only *one* new point. A most surprising one.

ROBB. Which point, Doctor?

OPPENHEIMER. That I strongly opposed the development of the hydrogen bomb, on moral and other grounds; that I turned other scientists against the hydrogen bomb; that I thereby considerably slowed down the development of the hydrogen bomb.

ROBB. In your opinion, Doctor, this allegation is not justified?

OPPENHEIMER. It is not true.

ROBB. Not true in any respect?

OPPENHEIMER. In no respect at all. Ever since our apprehensions concerning the monopoly of the hydrogen bomb have been proved right—ever since the two world powers have been facing each other like scorpions in a bottle—there have been people trying to persuade America that the blame lies with traitors.

ROBB. I would like, first of all, to deal with your former Communist associations, Doctor, taking the letter of the Atomic Energy Commission as the basis, and would like to have the letter set down in the record.

GARRISON. It would be appropriate for Dr. Oppenheimer's answering letter also to be set down in the record, Mr. Chairman.

GRAY. Very well, Mr. Garrison.

GARRISON. Furthermore, I would like to submit——

GRAY. Yes, please.

GARRISON. ——that allegations on which Dr. Oppenheimer was cleared in previous security investigations should not be the subject of the present investigation.

ROBB. Objection.

GRAY. Would you sustain your objection, Mr. Robb?

ROBB. The Atomic Energy Commission wishes to have certain allegations re-investigated, Mr. Chairman, on the basis of evidence not available at previous investigations.

MARKS. May I ask, Mr. Robb, what new evidence you wish to present, for instance, as to Point Three of the letter?

EVANS. Which point, Mr. Marks?

MARKS. Point Three. It says there that sixteen years ago, in 1938, Dr. Oppenheimer was an honorary member of the West Coast Council of the Consumers' Union. What new evidence is now available?

ROBB. There is fairly new evidence in the matter of a closed

Communist meeting at Dr. Oppenheimer's residence in
1941——

MARKS. I am asking you about Point Three——

ROBB. ——and a fairly new witness who has testified under
oath to what Dr. Oppenheimer thinks fit to deny.

MARKS. Does that witness happen to be Paul Crouch?

ROLANDER. Mr. Chairman, I would like to ask Mr. Marks why
he conjectures that the witness may be Paul Crouch.

MARKS. Paul Crouch figures rather excessively as a witness
these days, Mr. Rolander. No investigation of anybody's
loyalty without Paul Crouch, so to speak. It's his profession,
I guess.

ROLANDER. Mr. Chairman, I would like to ask Mr. Marks
whether, by some means or other, he has obtained information about Dr. Oppenheimer from secret F.B.I. files.

MARKS. No. Only you and Mr. Robb have such information.
That's the difference between an investigation and a trial.

EVANS. I beg your pardon, it is rather bewildering, I am not
used to this sort of thing. Mr. Rolander, who is this Paul
Crouch? I have never heard his name before.

ROLANDER. Paul Crouch is a former Communist functionary
who has turned his back on Communism.

EVANS. And he knows Dr. Oppenheimer?

MARKS. He knows Dr. Oppenheimer and he knows Malenkov,
but I guess they don't know him!

EVANS. It would have surprised me.

Slight laughter.

MARKS. I believe, Mr. Robb, you haven't yet answered my
question regarding Point Three.

ROBB. Indeed, I have not, Mr. Marks, as I uphold my objection for the following reasons: there is new evidence, and
there are new rules governing clearance; furthermore, it appears to me that there is a connection between Dr. Oppenheimer's former associations and his attitude in the matter
of the hydrogen bomb. I would therefore wish to retain the

right to question him, and other witnesses, as to these matters. In his own interest as well.

GRAY. Objection sustained.

Change of lighting. ROBB *steps forward to the footlights. The*
hangings close.

ROBB. People may think I am biased. They'd be wrong. When
I started, Oppenheimer was my idol among the scientists
of America, he was the Atom Bomb, he was—Oppie.

Then I studied his files—the material, four foot high,
which had led the F.B.I. to conclude that Oppenheimer
was "probably a camouflaged Soviet agent," and which
caused President Eisenhower to give the immediate order
to put "an impenetrable wall between Oppenheimer and
all government secrets." Those files changed the idol into a
sphinx.

Merits or no merits, we recently dismissed 105 officials
of the State Department for less incriminating associations
and less dangerous views. It was precisely in the most vital
sphere of nuclear energy that we came across the new type
of traitor—the traitor for ideological, ethical, and I don't
know what other motives. How could I be entirely sure
about Oppenheimer? I could not find the key to a number of contradictory facts in his life, I could not find the
key to his attitude in the matter of the hydrogen bomb.
But neither could I say: such and such facts are evidence
of his disloyalty. They remained debatable, alongside other
and equally debatable facts. I admit that particularly in the
case of Oppenheimer I have come to realize the inadequacy
of being strictly confined to facts in our modern security
investigations. How clumsy and unscientific is our proce-
dure when, over and above the facts, we do not concern
ourselves also with the thoughts, the feelings, the motives
which underlie those facts, and make them the subject of
our inquiries. It is the only method if we ever want to arrive
at a conclusive judgment as to Oppenheimer's integrity.

Do we dissect the smile of a sphinx with butchers' knives?
When the security of the free world depends on it, we must.

ROBB *returns to the stage itself.*

SCENE 2

The following text is projected on the hangings:

EXCERPT FROM THE PROCEEDINGS ON THE SECOND DAY:

GUILT THROUGH ASSOCIATION?

ROBB. Have you ever been a member of the Communist Party, Doctor?

OPPENHEIMER. No.

ROBB. Your wife?

OPPENHEIMER. At the time of her first marriage, yes. Up to 1936 or thereabouts.

ROBB. To whom was she married?

OPPENHEIMER. Joe Dallet.

ROBB. Was he a Communist?

OPPENHEIMER. He fell in the Spanish Civil War. I never knew him. . . . Yes.

ROBB. Was your brother Frank a member?

OPPENHEIMER. Until 1941.

ROBB. His wife, Jackie?

OPPENHEIMER. Yes.

ROBB. Was there a time when you yourself were in pretty close agreement with certain Communist ideas, Doctor?

OPPENHEIMER. Sure. I've put it all down in my answering letter.

ROLANDER *continues the examination.*

ROLANDER. On page five of your letter, you use the expression "fellow traveler." What exactly do you mean by it?

OPPENHEIMER. I would call a person a fellow traveler when he agrees with certain parts of the Communist program and when he is willing to co-operate with Communists without himself belonging to the Party.

ROLANDER. In the sense of your definition, sir, were you a fellow traveler?

OPPENHEIMER. Yes.

ROLANDER. When?

OPPENHEIMER. From about 1936 onward. After 1939 I did far less "fellow traveling," and after 1942 practically none at all.

ROLANDER. As from 1942, you would no longer call yourself a fellow traveler?

OPPENHEIMER. No. Some vague sympathies remained.

ROLANDER. How would you explain the fact that your sympathies cooled off so fast, precisely in 1942?

OPPENHEIMER. They had already cooled off considerably at the time of the purge trials under Stalin, and I had practically no sympathies left when there was that pact between the Nazis and the Russians. It made me sick when I heard that the gifted German physicist, Houterman, and a hundred other arrested German Communists had been handed over to the Gestapo by the Soviets.

ROLANDER. And your sympathies revived again, did they not, when Russia became our ally?

OPPENHEIMER. I guess we all felt pretty relieved.

ROLANDER. But in 1942, when you were put in charge of Los Alamos, your sympathies had once again cooled off?

OPPENHEIMER. What do you mean?

ROLANDER. I am trying to discover your motives, sir.

OPPENHEIMER. Motives? In what respect?

ROLANDER. You broke off your connections with a number of Communist friends, sir.

OPPENHEIMER. Yes, because I was working on the atom bomb! In the New Mexico desert, under military security restrictions. That way, all personal connections were broken off.

ROBB. Not all of them, I believe, Doctor. . . . Was your former fiancée, Dr. Jean Tatlock, a member of the Communist Party?

OPPENHEIMER. Yes. Less from political than from romantic

motives. She was a sensitive person, who profoundly despaired of the injustices in this world.

ROBB. How long was she a member of the Party?

OPPENHEIMER. It was an "on again, off again" affair. Right until her death, I think.

ROBB. How did she die, Doctor?

OPPENHEIMER [*after a pause*]. She killed herself. I believe F.B.I. agents have reported in detail how many days before that, and how long, I spent with her in such and such a hotel without informing the security authorities about our meeting.

ROBB. That is correct, Doctor. You spent the night with her, and——

OPPENHEIMER. What business is that of yours? What has it to do with my loyalty?

ROBB [*in a friendly tone*]. Has it nothing to do with your loyalty, Doctor, when *you*, the man responsible for the atomic weapons project in Los Alamos, when *you* spend the night in a hotel with a Communist woman—without informing the security authorities?

OPPENHEIMER. That Communist woman happened to be my former fiancée who was going through a severe emotional crisis and who wished to see me. A few days later, she was dead.

ROBB. What did you talk about, the two of you?

OPPENHEIMER. I do not propose to tell you.

ROBB. You won't tell me?

OPPENHEIMER. No.

He rises from the swivel chair, the "witness stand," and walks across to the sofa. He lights his pipe.

ROBB. Let the record show that Dr. Oppenheimer has left the stand as a witness.

GARRISON. Mr. Chairman, I object to this line of questioning. It is immaterial to the proceedings, and it encroaches upon Dr. Oppenheimer's privacy. The matter of the meeting with

Jean Tatlock has already been cleared up in previous security investigations.

GRAY. Objection sustained. . . . May I ask Dr. Oppenheimer to resume the stand as a witness.

OPPENHEIMER *resumes the stand.*

ROBB. The question was not put to you in an unfair spirit, Doctor.

OPPENHEIMER *glances at him, smoking his pipe. Change of lighting.* EVANS *steps forward to the footlights. The hangings close.*

EVANS. Perhaps I should have turned down this appointment, I probably should have, I am seventy years old. I cannot reconcile these interrogations with my idea of science; whose business are these private matters, what purpose do they serve, these humiliations? Is a humiliated man more loyal than a man who has not been humiliated? More devoted? It is now said in our universities: "Don't talk, don't write, don't move." If this continues, where is it going to lead us?

On the other hand, it was the physicists themselves who started the whole thing when they turned their profession into a military discipline, Oppenheimer in particular, Los Alamos was his idea. Take the dropping of the bomb, his explanations here. What more do they want? Do they need even greater submissiveness? I don't know, perhaps my liberal views are outmoded; perhaps science, too, must bow to the absolute claims of the state. Now that science has become so important. At any rate, I can see two kinds of development. The one is our increasing control over nature, our planet, other planets. The other is the state's increasing control over us, demanding our conformity. We develop instruments in order to pry into unknown solar systems, and the instruments will soon be used in electronic computers which reduce our friendships, our conversations, and thoughts to scientific data. To discover whether they are the right friendships, the right conversations, the right thoughts, which *conform*. But how can a thought be new, and at the same time conform? What is the difference between us and the dictatorships with their enforced conformity, if we go on as we do now? Perhaps I exaggerate. In one or two generations, scientists may take it for granted that they are functionaries. It makes me uneasy. I ask myself these questions as I listen to what is being said here. Is Oppenheimer only a beginning?

EVANS *returns to the stage itself.*

SCENE 3

The following text is projected on the hangings:

FROM THE PROCEEDINGS ON THE THIRD DAY:

ARE FORMER COMMUNIST SYMPATHIES COMPATIBLE WITH SECRET WAR PROJECTS?

ABOUT THE RELIABILITY OF PROFESSIONAL WITNESSES.

ROBB. Dr. Oppenheimer, yesterday you confirmed that, for a time, you were very closely associated with the Communist movement.

OPPENHEIMER. For a short time. Roughly, up to the end of the Spanish Civil War. Fifteen years ago.

ROBB. In those days, you attended meetings, trade union meetings, you had Communist friends, acquaintances, you belonged to a number of organizations with Communist leanings, you read Communist literature, you signed manifestoes, you paid out fairly substantial sums of money which went through Communist channels——

OPPENHEIMER. I gave money for the men who fought Franco and the Nazis in Spain. As you know, they depended on private support.

ROBB. You contributed up to three hundred dollars a month to the cause of the Spanish Republic and the money went through Communist channels?

OPPENHEIMER. If you had asked me for money to help that cause I'd have given it to you also.

ROBB. But you gave it to Isaac Folkoff, the Communist functionary, and you say in your answering letter to us, on page six, quote, "At that time, I agreed with the Communist idea that a Popular Front should be formed to oppose the spread of Fascism in the world." What does that mean?

OPPENHEIMER. It means that I was greatly perturbed about conditions in Germany and Spain, and that I did not want anything like that to happen here.

ROBB. What was it that perturbed you?

OPPENHEIMER. What perturbed me, Mr. Robb? That the world just looked on, hands in pockets. I had relatives in Germany, Jews, whom I was able to get across to this country, and they told me what was happening there.

ROBB. Quite so, Doctor, but didn't you know at the time that it was the tactics of the Communists to establish their rule everywhere by means of the so-called Popular Front?

OPPENHEIMER. Maybe it was their object. I myself did not see that danger. I saw what came spreading across the world from Germany, Italy, and Japan, and nobody did anything about it. That is how my sympathies started, and the manifestoes, and the donations. The best people of America put their names to the manifestoes. Those were different times.

ROBB. What I am getting at, Doctor, is this: as you were in such complete agreement with the Communists at that time, why did you not join the Party?

OPPENHEIMER. Because I don't like to think the thoughts of others. It goes against my idea of independence.

ROBB. Have you never thought of becoming a member?

OPPENHEIMER. No.

ROBB. Did your friends never suggest it to you?

OPPENHEIMER. No.

ROBB. What is your explanation for that?

OPPENHEIMER. They must have known me well enough. . . .

ROLANDER. Do you think it may possibly be Communist tactics, sir, to leave certain influential persons outside the Party because they would otherwise be less useful?

OPPENHEIMER. I don't know. I'm no expert.

ROLANDER. You do not regard yourself as very experienced in Communist affairs, sir?

OPPENHEIMER. No. By the time I started working on war projects, at Berkeley, my sympathies had almost completely cooled off.

ROBB. A closed Communist meeting takes place in a man's house. Does that point to "almost completely cooled-off sympathies," Doctor?

OPPENHEIMER. When is the meeting supposed to have taken place?

ROBB. So far, I did not speak of you.

OPPENHEIMER. I am sure you are referring to me.

ROBB. Since you are so sure: do you think it possible that on July 23, or thereabouts, in 1941, a closed Communist meeting was held in your house, at which a Communist functionary explained the new Party line?

OPPENHEIMER. No.

ROLANDER. Had you rented a house, 10 Kenilworth Court, in Berkeley, California, in July 1941, sir?

OPPENHEIMER. Yes.

ROLANDER. Do you know a man named Schneidermann?

OPPENHEIMER. Yes.

ROLANDER. Is he a Communist functionary?

OPPENHEIMER. Yes.

ROLANDER. How did you come to meet him?

OPPENHEIMER. I think I met him at Haakon Chevalier's house. At a literary meeting.

ROBB. Was Haakon Chevalier a frequent visitor at your house, in those days?

OPPENHEIMER. Yes.

ROBB. Was your pupil, Joseph Weinberg, a frequent visitor at your house, in those days?

OPPENHEIMER. Yes.

ROBB. It has been stated by two witnesses, Doctor, and the witnesses are prepared to testify under oath, that on July 23, or shortly after, you took part in a closed Communist meeting at 10 Kenilworth Court, in Berkeley, and that, at this meeting, Schneidermann explained the new line adopted by the Party as a consequence of Russia's entry into the war. According to the witnesses, there were present, among others: Haakon Chevalier, Joseph Weinberg, Dr. Oppenheimer, and Mrs. Oppenheimer.

OPPENHEIMER. That is not true.

ROLANDER. At that time, did you live in a Spanish-style bungalow which had colored wooden ceilings?

OPPENHEIMER. Yes.

ROLANDER. Was there a large candlestick of blue Venetian glass in your living room?

OPPENHEIMER. Yes.

ROLANDER. Was there a red wooden fairground horse standing by the fireplace?

OPPENHEIMER. Yes.

ROLANDER. These are some details of the furnishings as remembered by the witnesses. Maybe you have forgotten that meeting, sir?

MARKS. Mr. Chairman, may I ask Mr. Robb who these two witnesses are who remember candlesticks and fairground horses, and who wish to testify under oath that such a meeting took place?

ROBB. The witnesses are Paul Crouch and his wife.

EVANS. The same Crouch who has already been mentioned here?

ROBB. Yes, Dr. Evans.

MARKS. Mr. Chairman, I submit that the witnesses be called, to testify under oath.

ROLANDER. That is not possible, unfortunately.

GRAY. Why not?

ROLANDER. We ourselves would have liked to call the witnesses, Mr. Chairman, but the F.B.I. has not released them for our purposes here.

GRAY. I am sorry about that, Mr. Marks. Why did you want these witnesses called?

MARKS. I would have liked to show that their testimony is false, and that there are certain people who have a personal interest in that kind of testimony.

ROLANDER. Do you mean to say, sir, that the F.B.I. makes use of false testimony?

MARKS. That is not what I said. I do not know who the people are whose personal interests are involved here; I would have liked to question the witnesses about them. I only know that the testimony is false.

ROBB. I presume you will place your evidence for such knowledge before us, Mr. Marks.

MARKS. When is the meeting in Berkeley supposed to have taken place?

ROBB. In 1941. On July 23, or shortly after.

GARRISON. What exactly do you mean by that?

ROBB. Not before July 23, and not after July 30.

GARRISON. Did you have an opportunity to have this verified by the witnesses, Mr. Robb?

ROBB. Yes, recently.

MARKS *extracts a bundle of photostats from a file and takes it across to* GRAY.

MARKS. In that case, I would like to place evidence before the Board to the effect that from July 20 to August 10 Dr. Oppenheimer and his wife were not in Berkeley but in New Mexico. You will find here the names of the hotels where they stayed, and the names of the persons they met. I warned you about Paul Crouch, Mr. Robb. He isn't worth his witness fee.

ROBB. I see, Mr. Marks, that your office found it very important to prove Dr. Oppenheimer's absence at the time in question.

MARKS. Sure.

ROBB. Without previous knowledge of the allegation, I presume?

MARKS. We did it in the case of several longer journeys undertaken by Dr. Oppenheimer.

ROBB. I am perfectly convinced.

Change of lighting. MARKS *steps forward to the footlights. The hangings close.*

MARKS. Someday it will be realized, I hope, that it was not Oppenheimer who stood here before the Commission, but that it was our present-day security system which stood here. Oppenheimer is my friend; for many years, I have been General Counsel for the Atomic Energy Commission; I know the problems. If Oppenheimer is condemned here, our present-day security system will have passed judgment on itself, the subjugation of science to the military will have been proclaimed, and in their ranks there will be no room for independent spirits, for people who call a spade a spade.

If there had been no politics involved, if it had only been a matter of Oppenheimer himself, it would have been quite simple for the Atomic Energy Commission not to renew his contract—which expires in three months' time. Are the atomic secrets to vanish out of his head if his clearance is withdrawn? Lewis Strauss is the man whose first official action was to start these proceedings when he took over the Atomic Energy Commission—he is the same man who granted Oppenheimer's clearance in 1947. Now he cables the news of its suspension to the Air Force, the Army, and the Navy. Are these proceedings fair? The Board has access to secret F.B.I. files, but we are not allowed to see them. Oppenheimer cannot look at his own correspondence, his own reports; they have been confiscated and declared secret. I ask myself whether the course adopted by the defense is the right one; Oppenheimer wants us to follow this course, the defensive refutation of facts. But it is not a question of facts, or, if so, they are only a secondary matter. Why do we agree to have our battlefield here? Why don't we bring this issue out into the open, and place it before the whole body of scientists, before the public, whom it concerns? Are we waiting for the other side to make the attack—here, too? I do my utmost to convince Oppenheimer. His faith in the power of arguments makes him a worse witness than Joan of Arc, who could not read.

He returns to the stage itself.

SCENE 4

The following text is projected on the hangings:

FROM THE PROCEEDINGS ON THE FIFTH DAY:

WHERE DOES LOYALTY TO A BROTHER END, AND WHERE TO THE STATE?

SHALL A MAN BE PERSECUTED FOR HIS OPINIONS?

MORGAN. What I am interested in, Dr. Oppenheimer, is the practical side. Not the opinions but the consequences. You had to find the scientists for the project at Los Alamos, didn't you?

OPPENHEIMER. Yes. I suggested people whom I considered qualified. The final decision rested with General Groves and with Colonel Lansdale, the Chief of Security.

MORGAN. In your opinion, is it appropriate that a Communist should work on a secret war project?

OPPENHEIMER. When? In those days, or now?

MORGAN. Let us say, now.

OPPENHEIMER. As a rule, no.

MORGAN. And in those days?

OPPENHEIMER. In those days, an exception to the rule would have seemed more feasible to me.

MORGAN. Why?

OPPENHEIMER. In those days, Russia was our ally; now she is our potential enemy.

MORGAN. In other words, it is the Communist Party's connection with Russia that makes it impossible to have a Communist working on a secret war project?

OPPENHEIMER. Obviously.

MORGAN. When did this become obvious to you?

OPPENHEIMER. '46, '47.

MORGAN. Let me ask you a blunt question, Dr. Oppenheimer. Did you not know in 1943 that the Communist Party was an instrument of espionage in this country?

OPPENHEIMER. No.

MORGAN. You never suspected it either, in those days?

OPPENHEIMER. No. The Communist Party was legal. The Russians were our lauded allies who had just beaten Hitler at Stalingrad.

MORGAN. I never lauded them, I believe.

OPPENHEIMER. But you never tipped me off, either. Nor did you tip off the government.

MORGAN. How would you know that?

OPPENHEIMER. As far as the practical side is concerned, Mr. Morgan, nobody was employed at Los Alamos who was known to be a member of the Communist Party.

MORGAN. And you never suggested such a person, Dr. Oppenheimer?

OPPENHEIMER. No.

MORGAN. Why not?

OPPENHEIMER. Because of the problem of divided loyalty.

MORGAN. Divided between whom?

OPPENHEIMER. It seemed incompatible to me for a man to work, on the one hand, on secret war projects which, on the other hand, he is expected to destroy, according to the program of his Party.

MORGAN. I see.

ROBB. With reference to Los Alamos, Doctor, what dangers did you envisage in such collaboration?

OPPENHEIMER. The danger of indiscretion.

ROBB. Is that just another word for espionage?

OPPENHEIMER. It means less than that. But it does imply danger.

ROBB. At any rate, you regarded a Communist as too great a security risk?

OPPENHEIMER. An active Party member, yes.

ROBB. And what about former Party members? What did you do when you had to recommend a physicist who was a former Party member?

OPPENHEIMER. When I knew that, and thought he was dangerous as far as the secret war project was concerned, I recommended him but added my reservations.

ROBB. What test did you apply to determine whether a former Party member was still dangerous?

OPPENHEIMER. I gave my personal impression of him. . . . It was very difficult to find qualified people. We were working under extremely hard, extremely unpleasant conditions.

ROBB. You have not answered my question, Doctor.

OPPENHEIMER. Would you repeat your question?

ROBB. What test did you apply, in those days, to satisfy yourself that a former Party member was no longer dangerous?

OPPENHEIMER. What test? Applied to whom? To my wife?

ROBB. Let us take your brother, who is a physicist like yourself. Tell us about the test you applied to convince yourself that you could trust him.

OPPENHEIMER. In the case of a brother you don't apply tests. At least I didn't. I knew my brother.

ROBB. Well, what convinced you that your brother was no longer dangerous?

OPPENHEIMER. I never regarded my brother as dangerous. The possibility that a member of the Communist Party might engage in espionage never meant to me that *every* Party member would actually engage in espionage.

ROBB. I see. Your brother was an exception to the rule you have just told me about?

OPPENHEIMER. No. I did not say that every Communist is in fact a security risk, but that it would be a good policy to make that rule. Joliot-Curie in France is an example of the opposite. He is a Communist, and he is in charge of the French atomic weapons program.

ROBB. The atom spies, Klaus Fuchs, Nunn May, and Pontecorvo are examples of a different kind?

Photographs of the three men are projected.

OPPENHEIMER. Yes.

EVANS [*interested, turning to* OPPENHEIMER]. I beg your pardon, did you know Klaus Fuchs?

OPPENHEIMER. Not very well. He only came with the British Delegation to Los Alamos. He was connected with the Department of Theoretical Physics, of which Hans Bethe was in charge.

EVANS. What kind of a person was he?

OPPENHEIMER. Quiet, rather introvert, the son of a German pastor. He was crazy about driving cars, and drove with utter recklessness.

EVANS. I never understood his motives. Were they the usual kind of motives? Was he getting money from the Russians?

OPPENHEIMER. It seems he had rather pretentious ethical motives . . .

EVANS. Ethical motives? In what respect?

OPPENHEIMER. He told the British Secret Service he could not reconcile it with his conscience that the atom bomb should be left solely in the hands of a power which, he was afraid, might misuse the bomb. He rather fancied himself in the role of God, or the conscience of the world.

GRAY. Do such thoughts make any sense to you, Dr. Oppenheimer?

OPPENHEIMER. No. Not in that way.

EVANS. Do you think the Russians owe their atom bomb

mainly to the information they received from Fuchs, or May, or others?

OPPENHEIMER. Not essentially. They got to know that we were working on it. Also, certain details about our plutonium bomb. As far as I know, our secret services discovered that the Russians were following a different line of research. That is why Fuchs could not answer the questions they put to him—because of the difference in research.

ROBB. May I continue, Mr. Chairman?

GRAY. Indeed, yes.

ROBB. When did your brother become a member of the Communist Party?

OPPENHEIMER. 1936 or 1937.

ROBB. And when did he leave the Party?

OPPENHEIMER. In the fall of 1941, I believe.

ROBB. That was when he went from Stanford to Berkeley, to work in the Radiation Laboratory, is that right?

OPPENHEIMER. Yes. Lawrence wanted him there, for unclassified work.

ROBB. But shortly after that, he was doing work on secret war projects?

OPPENHEIMER. About a year later.

ROBB. After Pearl Harbor?

OPPENHEIMER. Possibly.

ROBB. Did you, thereupon, inform the security authorities that your brother had been a member of the Communist Party?

OPPENHEIMER. Nobody asked me.

ROBB. Nobody asked you. Did you tell Lawrence or anyone else about it?

OPPENHEIMER. I told Lawrence that my brother's troubles at Stanford arose from his Communist connections.

ROBB. Doctor, I didn't ask you quite that question. Did you tell Lawrence or anybody else that your brother, Frank, had actually been a member of the Communist Party?

OPPENHEIMER. No.

ROBB. Why not?

OPPENHEIMER. I don't think it is my duty to ruin my brother's career when I have complete confidence in him.

ROBB. How did you reach the conclusion that your brother was no longer a member of the Party?

OPPENHEIMER. He told me.

ROBB. That was enough for you?

OPPENHEIMER. Sure.

ROBB. Do you know that at that time, and also quite a while later, your brother publicly denied he had ever been a Party member?

OPPENHEIMER. I know he denied it in 1947.

ROBB. Why, in your opinion, did he deny it?

OPPENHEIMER. He probably wanted to go on working as a physicist—and not as a farmer, as he has been forced to do since then.

ROBB. Do you approve of his conduct, Doctor?

OPPENHEIMER. I don't approve of it. I understand it. I disapprove of a person being destroyed because of his past or present opinions. That is what I disapprove of.

ROBB. We are speaking about work on secret war projects and about the possibly disagreeable measures we have to take in order to protect our freedom, Doctor.

OPPENHEIMER. I know. There are people who are willing to protect freedom until there is nothing left of it.

ROBB. In the case of your brother, would I be right in saying that your quite natural loyalty to him outweighed your loyalty to our security authorities?

OPPENHEIMER. I have explained that there was no such conflict of loyalties.

ROBB. According to your own testimony, you were of the opinion that it would be important for the security authorities to know whether somebody had been a member of the

Communist Party. Yet, in the case of your brother, you concealed that fact from them, did you not?

OPPENHEIMER. I did not specifically conceal it. Nobody asked me.

ROBB. And you did not volunteer this information?

OPPENHEIMER. No.

ROBB. That is all I wanted to know, Doctor.

Change of lighting. ROLANDER *steps forward to the footlights. The hangings close.*

ROLANDER. People argue that we judge past events from our present point of view. Yes, that is so, for we are investigating whether Dr. Oppenheimer is a security risk *today*, when our enemies are the Communists, Russia, and not the Nazis, as in the past. Facts are very relative. For instance, in 1943 we certainly would not have told our most vital secrets to a man whose sympathies were with the Nazis, even if he was a genius; and now, in 1954, this is our attitude towards a man whose sympathies are with the Communists. Security measures are *realistic*: what has to be protected against whom, and in what situation. They have no claim to absolute justice and immaculate morality. They are *practical*. That is why I am disturbed by these ideological exercises here, this flogging of principles about the sacredness of privacy, a thing that dates back to the last century. We must examine quite soberly how strong Oppenheimer's sympathies were, how persistent they are, what consequences this had for us in the past, and whether we can afford such consequences in the future. . . . It is history itself—the possibility of the free world being destroyed—which makes our security measures rigorous and uncompromising.

I feel so old among these people who are older than me. Where they have their ideology I have a blind spot.

He returns to the stage itself.

SCENE 5

The following text is projected on the hangings:

FROM THE PROCEEDINGS ON THE SEVENTH DAY:

WHAT KIND OF PEOPLE ARE PHYSICISTS?

CAN A MAN BE TAKEN TO PIECES LIKE THE
MECHANISM OF A FUSE?

ROLANDER. Speaking of secret war projects, sir, do you agree that, with a fellow traveler, there is a potentially greater danger of indiscretion?

OPPENHEIMER. Potentially, yes. It depends on the person.

ROLANDER. Is it a fact, Dr. Oppenheimer, that a considerable number of scientists in Los Alamos were fellow travelers?

OPPENHEIMER. Not very many. Less than at Berkeley, for instance. But in those days we'd have snatched a man from the electric chair if we needed him to get the thing going.

ROLANDER. What I cannot understand, sir, is this: why exactly was it the fellow travelers who were snatched from the electric chair in such numbers?

OPPENHEIMER. Because there were many physicists with left-wing views.

ROLANDER. How would you explain that?

OPPENHEIMER. Physicists are interested in new things. They like to experiment and their thoughts are directed toward changes. In their work and also in political matters.

ROLANDER. Many of your pupils were in fact Communists or fellow travelers, were they not?

OPPENHEIMER. A few of them, yes.

ROLANDER. Weinberg, Bohm, Lomanitz, Friedman?

OPPENHEIMER. Yes.

ROLANDER. And you recommended these young men for work at Berkeley or Los Alamos?

OPPENHEIMER. I recommended them as scientists, yes . . . Because they were good.

ROLANDER. Purely professionally. I see.

OPPENHEIMER. Yes.

ROLANDER. Many of your intimate friends and acquaintances —professional and otherwise—were also fellow travelers, were they not?

OPPENHEIMER. Yes. I don't find that unnatural. There was a time when the Soviet experiment appealed to all those who were dissatisfied with the state of the world which, I would agree, is certainly unsatisfactory. But now we view the Soviet experiment without any illusions; now Russia is facing us as a hostile world power. And so we condemn the hopes which many people had set on that experiment. . . . It seems to me unwise to condemn those hopes, and I feel it is wrong that people should be disparaged and persecuted because of their views.

ROLANDER. I do not wish to disparage anybody, sir. I merely pursue the question whether a physicist is a greater security risk when he has a certain number of friends and acquaintances who are former Communists or fellow travelers. Is he not in fact a greater security risk?

OPPENHEIMER. No.

ROLANDER. You think it is immaterial, nowadays too, how many acquaintances with Communist leanings——

OPPENHEIMER. I think that a man cannot be taken to pieces like a fuse mechanism. Such and such views: such and such security. Such a number of acquaintances who are fellow travelers: such a degree of security. These mechanical calculations are folly. If we had proceeded like that in Los Alamos, we'd never have employed the best people. We might perhaps have had a laboratory full of men with the most irreproachable views in the world, but I don't believe the thing would have worked. People with first-class ideas don't pursue a course quite as straight as security officers fondly imagine. You cannot produce an atomic bomb with irreproachable, that is, conformist ideas. Yes-men are convenient, but ineffectual.

ROLANDER. What did you do, sir, when you heard in 1947 that Weinberg, Bohm, and others, were active Party members?

OPPENHEIMER. What do you mean?

ROLANDER. Did you break off all connections with them?

OPPENHEIMER. No.

ROLANDER. Why not?

OPPENHEIMER. Because it isn't exactly my idea of good manners.

ROLANDER. Is it your idea of security?

OPPENHEIMER. What?

ROLANDER. Did you recommend your own lawyer to Weinberg, sir?

OPPENHEIMER. It was my brother's lawyer, I believe.

ROLANDER. Did you give a party for Bohm?

OPPENHEIMER. I went to a farewell party given for him when he had been fired at Princeton and was going to Brazil.

ROLANDER. And you found it quite easy to reconcile such a demonstration of sympathy for active Communists with the duties of your high office, as the Adviser to the Government on atomic matters?

OPPENHEIMER. What has that to do with atomic matters? I gave some advice to old friends, and I said good-by to them.

ROLANDER. Would you do the same again, today?

OPPENHEIMER. I should hope so.

ROLANDER. Thank you, sir.

GRAY. Any further questions to Dr. Oppenheimer?

Evans raises his hand.

EVANS. I am surprised that there were actually so many physicists who were Reds. Perhaps it depends on the generation.

OPPENHEIMER. I would say they were Pink, not Red.

EVANS. I can't understand why these people, otherwise quite sensible, were so strongly attracted to radical political ideas. What kind of people are physicists?

OPPENHEIMER. You think they might be a bit crazy?

EVANS. I have no idea, maybe eccentric; how do they differ from other people?

OPPENHEIMER. I think they simply don't have so many preconceived notions. They want to probe into things that don't work.

EVANS. I have never read Marx and such people. I have never taken an interest in politics, as you obviously have.

OPPENHEIMER. I wasn't interested either. For a long time. In my childhood, nobody prepared me for such bitter and cruel things as I got to know later, during the great depression, when my students went hungry and couldn't find jobs, the same as millions of others. I realized that something is wrong with a world in which such things can happen. I wanted to discover the reasons for that.

EVANS. And that is why you read Communist books in those days, sociology, and that sort of thing?

OPPENHEIMER. Yes. Although I never understood Marx's *Kapital,* for instance. I never got beyond the first fifty pages.

EVANS. Oddly enough, I have never met anybody yet who understood it. Apart from Rockefeller, perhaps . . . [*Laughter from* MORGAN, MARKS, *and* ROBB.] . . . and my dentist who says every time he drills into a nerve: "Marx tells us this, and Marx tells us that." [*Laughter.*] To me, therefore, Marx is always associated with a specific pain in a nerve. [*Laughter.*]

OPPENHEIMER. That seems to be the case with many people.

EVANS [*laughing*]. Of all the known philosophers whose works are never read, he is the one who causes us most trouble. Look where it's got *you!*

OPPENHEIMER *laughs. The lighting changes.* MORGAN *steps forward to the footlights. The hangings close.*

MORGAN. I had a talk with Gray yesterday. He was very annoyed because the Secretary of War has intervened. And has stirred up the scientists, of course. Playing with fire. I said I felt that there was a bit too much general discussion of Oppenheimer's political background and opinions. That may be just the thing for McCarthy, but not for those complicated highbrows, the physicists. We should make it clear to the scientists that we don't dictate such and such opinions to them, and that we don't intend to boot them out because they hold this or that opinion. But we must insist on a sharp dividing line between their subjective views and their objective work, because modern nuclear policy is possible only on that basis. This applies not only to science but also to industrial enterprise, to the modern state. That is why this Commission should not rest content with a documentation of Oppenheimer's political background, however astonishing it may be. Rather, we should find out whether, to our cost, he has wrongfully allowed that background, and those political, philosophical, moral views, to affect his work as a physicist and a government adviser. And whether this might be feared in the future. It is this aspect of the matter which makes his clearance an acute question, and this should be understood both by the public and by the physicists. No matter how extreme, the subjective views of a physicist are his own private affair as long as they don't interfere with his objective work. This dividing line bears upon the principles of our democracy.

He returns to the stage itself.

SCENE 6

The following text is projected on the hangings:

FROM THE PROCEEDINGS ON THE TENTH DAY:

WHAT IS ABSOLUTE LOYALTY?

IS THERE SUCH A THING AS HUNDRED-PER-CENT SECURITY?

Robb. I see from my files that today is your fiftieth birthday, Doctor. I should like, for a moment, to lay aside the formality of these proceedings and wish you many happy returns.

OPPENHEIMER. Thank you. There is no need.

ROBB. May I ask, Doctor, whether you have already seen your birthday mail?

OPPENHEIMER. Some of it.

ROBB. Has Haakon Chevalier written to you?

OPPENHEIMER [*with a short contemptuous laugh*]. Yes, a greeting card.

ROBB. What does he write?

OPPENHEIMER. The usual good wishes. "In old friendship, ever yours, Haakon." No doubt you have a photostat copy of it.

ROBB [*smiling*]. You still regard him as your friend, don't you?

OPPENHEIMER. Yes.

ROBB. In your answering letter to the Atomic Energy Commission, on page twenty-two, you recount a conversation you had with Chevalier in the winter of 1942–43. Where did that conversation take place?

OPPENHEIMER. At my house, in Berkeley.

EVANS. I beg your pardon, I'd just like to know a little more about this Chevalier; who was he, what kind of a person was he?

OPPENHEIMER. He was a member of the faculty.

EVANS. A physicist?

OPPENHEIMER. No. French literature.

EVANS. A Communist?

OPPENHEIMER. He has strong left-wing views.

EVANS. Red, or Pink?

OPPENHEIMER. Pinkish-red.

EVANS. And as a person?

OPPENHEIMER. One of the two or three friends one has in a lifetime.

ROBB. In your letter, you give the gist of the conversation. Now I would like to ask you, Doctor, to tell us about the circumstances and, if possible, to give us a verbatim account of that conversation.

OPPENHEIMER. I can only give you the substance, not the exact wording. It is one of those things I have too often thought about . . . eleven years ago. . . .

ROBB. Very well, then.

OPPENHEIMER. One day, in the evening, Chevalier came to our house, with his wife, I think he came for dinner, or for drinks——

GRAY. Excuse me, was it he who had made the appointment?

OPPENHEIMER. I don't know. It was simply that one of us called up the other, and said: "Why not drop in."

GRAY. I think it is important, Dr. Oppenheimer, that you should give us the story as much in detail as possible.

OPPENHEIMER. All right. . . . They came to our house, we had a brandy, we talked about the latest news, possibly about Stalingrad—it was about that time. . . .

GRAY. Did Chevalier introduce the subject of Stalingrad?

OPPENHEIMER. I don't know, it may be that we talked about it on some other day, but I think it was on that particular evening. One thing is certain, anyway: when I went into the kitchen to mix some drinks Chevalier followed me there and told me that he had recently met Eltenton.

GRAY. Would you tell us, for the record, who Eltenton is?

OPPENHEIMER. A chemical engineer, an Englishman, who had been working in Russia for several years.

GRAY. Party member?

OPPENHEIMER. Closely associated. Whether he actually was a member?—I didn't know him all that well.

ROBB. What did Chevalier want from you?

OPPENHEIMER. I'm not sure whether he *wanted* anything. He told me that Eltenton was furious because we had left the Russians in the lurch, because we weren't opening up a second front, and didn't give them the technical information they needed, and that it was a bloody disgrace.

ROBB. Was this Chevalier's opinion?

OPPENHEIMER. He spoke about Eltenton. He said Eltenton had told him he knew certain ways and means of transmitting technical information to Soviet scientists.

ROBB. What ways and what means, Doctor?

OPPENHEIMER. Chevalier did not say what they were. I don't know whether Eltenton had said what they were. We did not discuss things, I mean, I just said: "But that is treason!" ... I'm not sure, but, anyway, I said something to the effect that it was horrible and quite unspeakable. And Chevalier said he entirely agreed with me.

ROBB. Is that all that was said?

OPPENHEIMER. Then we talked about drinks, and about Malraux, I believe.

ROBB. Would you spell that name, Doctor?

OPPENHEIMER. M-a-l-r-a-u-x.

ROBB. Malraux? Who is he?

OPPENHEIMER. A French author. Chevalier translated his books.

ROBB. Is Malraux a Communist?

OPPENHEIMER. He used to be. Now he is the brain of de Gaulle.

ROBB. I have never heard his name.

GRAY. Did Chevalier know that, at Berkeley, work was being done on the development of the atomic bomb?

OPPENHEIMER. No.

ROBB. Did you use the word "treason," Doctor?

OPPENHEIMER. It is such a hackneyed word, I could tell you the whole history of the word "treason."

ROBB. Would you first answer my question?

OPPENHEIMER. I don't know.

ROBB. Did you regard it as treason?

OPPENHEIMER. What?

ROBB. To transmit secret information to the Russians?

OPPENHEIMER. Of course.

ROLANDER. Did you, thereupon, report the incident to your security authorities, sir?

OPPENHEIMER. No.

ROLANDER. Why not?

OPPENHEIMER. I didn't take that conversation very seriously. Just the way people talk, at parties.

ROBB. But six months later, Doctor, you took that conversation so seriously that you specially went from Los Alamos to Berkeley in order to draw the attention of the security authorities to it. Why?

OPPENHEIMER. Lansdale, the Chief Security Officer, had been in Los Alamos, and he had told me he was very worried about the situation at Berkeley.

ROBB. Do you agree with me, Doctor, when I say that his remark implied a fear of espionage?

OPPENHEIMER. That is right.

ROLANDER. Did he mention any names?

OPPENHEIMER. Lomanitz came up in the conversation. He had gossiped to people about things that were none of their business.

ROLANDER. Which people?

OPPENHEIMER. Of the C.I.O. union; that is why Eltenton came to my mind. Eltenton was rather active in the Scientists' and Engineers' Union.

ROBB. Did you tell Lansdale that you regarded Eltenton as a possible danger?

OPPENHEIMER. I first hinted at it to Johnson, the local Security Officer at Berkeley.

ROBB. Did you tell Johnson the story exactly as it had happened?

OPPENHEIMER. No, I said little more than that Eltenton was somebody not to be trusted. He asked me why I said this. Then I invented a cock-and-bull story.

ROBB. You lied to him?

OPPENHEIMER. Yes. I thought that would be the end of the

matter. But Johnson informed Radzi, his superior. Then I had an interview with both of them.

Robb. Did you tell Radzi the truth?

Oppenheimer. I told him the same story, but more in detail.

Robb. What in the story was not true?

Oppenheimer. That Eltenton had attempted to approach *three* members of the project, through intermediaries.

Robb. Intermediaries?

Oppenheimer. Or through *one* intermediary.

Robb. Did you disclose to Radzi the identity of the intermediary, Chevalier?

Oppenheimer. I only disclosed the identity of Eltenton.

Robb. Why?

Oppenheimer. I wanted to keep Chevalier out of it, and myself too.

Robb. In that case, why did you encumber him with contacting *three* members of the project?

Oppenheimer. Because I was an idiot.

Robb. Is that your only explanation, Doctor? [Oppenheimer *makes a gesture.*] Didn't you realize that Radzi and Lansdale would move heaven and earth to discover the identity of the intermediary and the three members of the project?

Oppenheimer. I ought to have known.

Robb. And didn't they move heaven and earth?

Oppenheimer. Yes. In the end, I promised Lansdale I would disclose the names if General Groves made that a military order. When Groves then commanded me to do so, I named Chevalier and myself.

Robb. That is all. [*Addressing himself to* Gray.] I would now like to hear Major Radzi as a witness.

Rolander. When you told Major Radzi that "cock-and-bull story," sir, there was also something about microfilms, wasn't there, and about a man in the Russian Consulate?

Oppenheimer. I can't imagine. No.

Rolander. Thank you, sir.

GRAY. Now we have two witnesses, Major Radzi and Mr. Lansdale. Major Radzi is a witness called by Mr. Robb, and will be questioned first. [*An* OFFICIAL *ushers* MAJOR NICHOLAS RADZI *in, by the door to the right, and leads him to the witness stand.* RADZI *is in mufti. He bows slightly to the members of the Board.*] Nicholas Radzi, do you swear that the testimony you are to give the Board shall be the truth, the whole truth, and nothing but the truth, so help you God?

RADZI. I do. [*He seats himself.*]

GRAY. A few questions regarding your personal history. What do you specialize in?

RADZI. Counterespionage in relation to war projects. In particular, countering Communist agents.

GRAY. How long have you been engaged in that sort of work?

RADZI. Fourteen years.

GRAY. Did you receive special training for it?

RADZI. I received the kind of training the F.B.I. give *their* top people. Rather rigorous training. Since then, I've had some international experience.

GRAY. Could you tell us about one of your special assignments?

RADZI. Towards the end of 1943, I and my group had to find out whether the Germans were developing an atomic bomb. And I had to kidnap the German eggheads in question—before they were kidnaped by the Russians. I guess we did that rather well.

GRAY. Were you specially trained how to deal with scientists?

RADZI. Yes. I think I have a sort of knack for it. By now, I have a pretty shrewd idea how a nuclear physicist's mind works, and how to get at him.

GARRISON. Mr. Chairman, may I ask, for the record: what was the original profession of Mr. Radzi?

RADZI. Physical training instructor. [*Laughing.*] I was a natural boxer and quite a good football coach.

GARRISON. Mr. Chairman, may I ask Mr. Radzi whether it was his own wish to appear here as a witness?

GRAY. Major Radzi?

RADZI. No. I have been ordered here by my department.

GRAY. Well, you know of course, Major, that you must give nothing but your own testimony here, and that you are not to follow instructions you might possibly have been given. . . . You may question the witness, Mr. Robb.

ROBB. I would like to ask you, Major, what particular assignment brought you into contact with Dr. Oppenheimer in 1943?

RADZI. Well, yes, in May 1943 I was asked to investigate a possible case of espionage at Berkeley. We only knew that a man named Steve Nelson, a prominent Communist functionary in California, had attempted to secure information concerning the Radiation Laboratory. And that he had tried to get this information through a man of whom we only knew that his first name, or his code name, was Joe— and that Joe had come from New York, and had sisters living in New York. We started the investigation, and at first we thought the man was Lomanitz. That is why we wanted to remove him from the laboratory and push him off into the Army.

ROBB. But you were not able to do that? Why?

RADZI. Oppenheimer used his personal connections to keep Lomanitz at Berkeley. Then it turned out that Lomanitz was not identifiable with Joe. For a while, we thought it might be David Bohm. And, later, Max Friedman. Finally we discovered that Joe was Joseph Weinberg.

ROBB. What had that investigation to do with Dr. Oppenheimer?

RADZI. It seemed somewhat strange that everybody we suspected had some sort of connection with Dr. Oppenheimer. Whenever we trod on somebody's toes he was sure to turn to Dr. Oppenheimer for help.

ROBB. Thereupon, what steps did you take?

RADZI. We asked the F.B.I., in June 1943, to place Dr. Oppenheimer under investigation because of suspected espionage.

ROBB. Did you conduct that investigation?

RADZI. Yes.

ROBB. What did you discover?

RADZI. That Dr. Oppenheimer had probably been a member
of the Communist Party and that he still had strong Communist leanings. Also, that he had associations with Communists such as David Hawkins and Jean Tatlock who, in
turn, had contacts with Steve Nelson and, through Steve
Nelson, possibly also with the Russians.

ROBB. Thereupon, what steps did you take, Major?

RADZI. We recommended to the Pentagon—to Mr. Lansdale,
my superior—that Dr. Oppenheimer should be completely
removed from any employment by the United States government. In case Dr. Oppenheimer was regarded as irre-
placeable, we recommended them to use the pretext that he
was endangered by certain Axis agents, whereupon he
should have two bodyguards assigned to him—men specially
trained by our department, who would watch him all the
time. Much to our regret, neither Lansdale nor Groves
accepted these recommendations.

ROBB. That happened quite some time before you had an
interview with Dr. Oppenheimer, is that right?

RADZI. Two months before.

ROBB. In that case, did it not surprise you that Dr. Oppenheimer himself, in August, volunteered to give you some
information about suspected espionage?

RADZI. Not particularly. It is quite a well-known reaction from
people who have discovered that their personal affairs are
being investigated.

ROBB. Do you recall your interview with Dr. Oppenheimer?

RADZI. Sure. I listened to the tape yesterday. . . .We made
a recording of it in Lieutenant Johnson's office, at the time.
[*He extracts a reel of tape from his briefcase.*] Here it is.

ROBB. May we hear it?

RADZI. Sure. It has been released by the F.B.I. [*He hands the tape to an* OFFICIAL *who fits it into a tape recorder.*]

ROBB [*to the* OFFICIAL]. Ready? Well, let us start.

The OFFICIAL *switches on the tape recorder. Projected on the screens: photographs of* RADZI *in summer uniform, and of* OPPENHEIMER, *1943, young-looking, sunburned, in shirt and slacks.*

Alternative suggestion: the interview RADZI-OPPENHEIMER-JOHNSON, *on an 8-mm. film, is projected, reasonably synchronized with the tape recording. The interview takes place on a hot August day at Los Alamos, in an office in an army hut.* OPPENHEIMER *in shirt and blue jeans, the officers in summer uniform.* JOHNSON, *seated at a small desk, surreptitiously manipulates a tape recorder.* RADZI *sees to it that the microphone is close to* OPPENHEIMER. *The whole film should look rather the worse for wear, in order to look like a genuine documentary. On no account must it give the impression of being a sound film.*

Tape recording:

RADZI. I am delighted, Dr. Oppenheimer, to meet you at last, and to be able to talk with you.

OPPENHEIMER. The pleasure is mine, Major.

RADZI. No, no, Doctor, on the contrary—you are one of the most important men in the world today, one of the most fascinating men, there's no doubt about it. Whereas we are only something like night watchmen. [*He laughs.*] I don't want to take up much of your precious time . . .

OPPENHEIMER. That's perfectly all right. Whatever time you want.

RADZI. . . . but Lieutenant Johnson told me yesterday you thought it possible that a certain group is interested in the project. He said you had been kind enough to give us this tip.

OPPENHEIMER. Yes, well, it is quite a while ago, and I have no first-hand knowledge, but I think it is true to say that a man, whose name I never heard, who was attached to the

Soviet Consulate, has tried to indicate through intermediaries to people concerned in the project that he was in a position to transmit information, with no danger to anybody.

RADZI. Information for the Russians?

OPPENHEIMER. Yes. We all know how difficult the relations are between the two allies, and there are many people—those, too, who don't feel particularly friendly towards Russia—who think it is wrong that we deny the Russians certain technical information, radar, and so on, while they battle for their lives, fighting the Nazis.

RADZI. Yes, it sure is a problem. . . . Perhaps you know that I have Russian blood.

OPPENHEIMER. There may be some arguments for giving the Russians official information, but it is obviously quite out of the question to have such information moving out the back door.

RADZI. Could you give us a little more specific information as to how these approaches were made?

OPPENHEIMER. They were made quite indirectly, well, I know of two or three cases. Two or three of the people are with me at Los Alamos, and they are closely associated with me. That is why I feel I should give you only *one* name, which has been mentioned a couple of times, and this person may possibly be an intermediary. His name is Eltenton.

RADZI. Eltenton? Does he work on the project?

OPPENHEIMER. No. He is employed by the Shell Development Company. At least, he used to work there.

RADZI. Were these people contacted by Eltenton himself?

OPPENHEIMER. No.

RADZI. Through another party?

OPPENHEIMER. Yes.

RADZI. Would you tell us, please, through whom these contacts were made?

OPPENHEIMER. I don't think that would be right. I don't want to have people involved who have nothing to do with the whole thing. It wouldn't be fair. They confided in me, and they have been one-hundred-per-cent loyal. It's a question of trust.

RADZI. Obviously, we don't mistrust these people, Doctor— just as little as we mistrust *you*. [*He laughs.*] It would be absurd! But we must have the intermediary, so that we can get inside the network.

OPPENHEIMER. I wouldn't like to give you his name. I have implicit confidence in him. . . . But it is a very different matter when Eltenton comes and says he has good contacts with a man attached to the Russian Consulate, who has a lot of experience with microfilms or whatever the hell.

RADZI. Of course I am trying to get out of you everything I possibly can. Once we've licked blood—we bloodhounds [*He laughs.*]—we are persistent.

OPPENHEIMER. It is your duty to be persistent.

RADZI. At any rate, I'm glad you have such a positive attitude to our work. Certain scientists don't find that so easy.

OPPENHEIMER. Whom are you referring to?

RADZI [*laughs*]. I'm not referring to Niels Bohr—[*He laughs.*] —to whom I explained for three solid hours what he must *not* say. And then he said the whole lot within the first half-hour of a railway journey. [OPPENHEIMER *laughs.*] When we got him across from Denmark we had to drag him out of the plane, unconscious, because he had forgotten to work his oxygen mask which we had made him put on. We flew at an altitude of 36,000 feet. [*He laughs.*]

   The OFFICIAL *switches off the tape recorder.*

EVANS [*turning to* OPPENHEIMER]. Was Niels Bohr with you at Los Alamos?

OPPENHEIMER. For a short time. Under a code name, like all of us. Nicolas Baker. He did not want to stay.

EVANS. Why not?

OPPENHEIMER. He was furious with us. He said we were turning science into an appendage of the military, and the moment we gave an atomic cudgel into their hands they'd let fly with it. It worried him a great deal.

Evans. He was the most charming person I've ever known.

Rolander. Doesn't it emerge quite clearly from the tape recording, sir, that you spoke to Major Radzi of "a man in the Russian Consulate who had a lot of experience with microfilms"?

Garrison. The exact words were, "with microfilms or whatever the hell."

Rolander. If you have a friend, sir, whom you regard as innocent, and you wish to protect him from the security authorities—why do you encumber him with the Russian Consulate, microfilms, and three contacts? I just can't see the reason for it.

Oppenheimer. Neither can I.

Rolander. You have no explanation?

Oppenheimer. Not an explanation that would sound logical.

Robb. Do you have an explanation for Dr. Oppenheimer's behavior, Major Radzi?

Radzi. Sure. Dr. Oppenheimer told us the truth, at the time.

Robb. You think that the story, which Dr. Oppenheimer called a cock-and-bull story here, is in fact true?

Radzi. Yes. And I think that his later attempt at minimizing it was the real cock-and-bull story. Dr. Oppenheimer told us about three genuine cases of people being contacted. This way, he wanted to secure our continued confidence in him, because he was afraid we might discover these contacts in the course of our investigations. But when our investigations weren't very successful he made light of the whole thing.

Robb. Was that your opinion also at the time?

Radzi. Sure. I told Lansdale all about it.

Robb. And Lansdale?

Radzi. The whole thing went up in blue smoke when Dr. Oppenheimer named Chevalier and himself. There was some more investigating, but finally the whole lot was thrown overboard.

Robb. From your knowledge of F.B.I. files, and from your own experience in this matter, would you grant Dr. Oppenheimer his security clearance?

RADZI. I wouldn't have cleared him then, and I wouldn't clear him now.

ROBB. Were you alone in that opinion?

RADZI. I guess all security officers below the rank of Lansdale and General Groves shared that opinion.

ROBB. Thank you very much, Major.

GRAY [*to* OPPENHEIMER'S *counsel*]. Would you wish to cross-examine Major Radzi?

MARKS. Yes . . . I am going to ask you a psychological question, Mr. Radzi. Is Dr. Oppenheimer's personality easy to understand, or is it rather complicated?

RADZI. Extremely complicated. And extremely contradictory.

MARKS. In other words: one would have to know him very well in order to reach a conclusive judgment?

RADZI. Yes.

MARKS. How well do you know Dr. Oppenheimer?

RADZI. I know him very well, in so far as I know his file very well.

MARKS. How often have you talked with him?

RADZI. Once.

MARKS. Does one get to know a person better from his file or from actual conversation with him?

RADZI. In our work, I would give preference to files. They are the sum total of all the impressions of a man which one single person can't get, just by himself.

MARKS. How long have Dr. Oppenheimer's activities been under surveillance by the security authorities, in particular by the F.B.I.?

RADZI. Thirteen or fourteen years.

MARKS. During that time, was there any proof of Dr. Oppenheimer being indiscreet?

RADZI. No proof.

MARKS. Or disloyal?

RADZI. In the Chevalier business, there is no doubt that Dr. Oppenheimer's loyalty to a friend outweighed his loyalty to America.

MARKS. Did Chevalier prove to be innocent?

RADZI. It was not possible to prove that he was guilty.

MARKS. What happened to him?

RADZI. He was fired at Berkeley and he was of course put under surveillance.

MARKS. If Dr. Oppenheimer foresaw such consequences, would it not be understandable that he hesitated several weeks before naming him?

RADZI. No, not when the safety of the country is at stake. From a scientist of such stature, we must demand absolute loyalty.

MARKS. Do you know that, in 1946, the F.B.I. investigated the Chevalier incident a second time?

RADZI. Yes.

MARKS. And that Mr. Hoover, the F.B.I. Chief, took an interest in it himself?

RADZI. Yes.

MARKS. And that Dr. Oppenheimer's clearance was granted, with no reservations?

RADZI. I'd have liked to see the man who'd have cast doubt on Dr. Oppenheimer's clearance in 1946, what with his prestige and influence! He was a god in those days.

MARKS. I have no more questions.

GRAY. Any further questions to Major Radzi? [EVANS *indicates his wish to speak.*] Dr. Evans.

EVANS. What I have always wanted to hear from an expert, it interests me, is somewhat general. In your opinion, Mr. Radzi, with reference to a secret war project, is it possible to achieve one-hundred-per-cent security?

RADZI. No. It would be possible to have ninety-five-per-cent security if scientists and technicians were selected with proper care, and if they were taught some understanding of our problems.

EVANS. What do you mean?

RADZI. They should be made to realize that, nowadays, they are experts working within one vast enterprise. They have to do their own particular share of the work and then hand it over to the other experts, the politicians and the military, who then decide what is to be done with it. And we are the experts who make sure that there are no rubbernecks looking in. If we want to defend our freedom successfully, we must be prepared to forego some of our personal liberty.

EVANS. I don't know, it doesn't make me feel too good, but it was interesting to hear the views of an expert.

GRAY. Any further questions? Mr. Morgan.

MORGAN. Do you think that Dr. Oppenheimer's Communist sympathies had something to do with his attitude in the Chevalier incident?

RADZI. Beyond a doubt. Although I have come to the conclusion that Dr. Oppenheimer can give his undivided loyalty to only two things: science and his own career.

GARRISON. Do you regard Mr. Lansdale, your former superior, as an incompetent security expert, Mr. Radzi?

RADZI. No. He is the most competent amateur I've ever known. Perhaps he lacks toughness, which is vital in our business.

GRAY. If there are no further questions—may I thank you for having come here to help us. [RADZI *gets up and leaves the room.*] Would Mr. Lansdale now be asked to appear?

An OFFICIAL *leaves the room to fetch* LANSDALE.

ROBB. To complete the record, Dr. Oppenheimer. You maintained your good relations with Chevalier, is that right?

OPPENHEIMER. Yes.

ROBB. When did you last see him?

OPPENHEIMER. A few months ago, in Paris.

ROBB. Doctor, when did your friend, Haakon Chevalier, first discover that it was you who had reported his case to the security authorities?

OPPENHEIMER. I guess he will discover it now, from these proceedings.

EVANS. You have never told him, I mean, this is almost private, that you set the whole thing going?

OPPENHEIMER. No.

EVANS. Why not?

OPPENHEIMER. I guess he wouldn't have understood.

The OFFICIAL *who went to fetch* LANSDALE *opens the door and looks questioningly at* GRAY.

GRAY. Is Mr. Lansdale there? [*The* OFFICIAL *leads* LANSDALE *to the witness stand.*] Do you wish to testify under oath, Mr. Lansdale?

LANSDALE. I leave that to counsel or to the Board.

GRAY. The previous witnesses have testified under oath.

LANSDALE. Then let us keep it uniform.

GRAY. John Lansdale, do you swear that the testimony you are to give the Board shall be the truth, the whole truth, and nothing but the truth, so help you God?

LANSDALE. I do.

GRAY. Will you please take the stand. . . . You are at present working as a lawyer, Mr. Lansdale?

LANSDALE. Yes, in Ohio.

GRAY. Where did you study?

LANSDALE. Harvard.

GRAY. You were responsible for the security of the entire atomic weapons project, were you not?

LANSDALE. During the war.

GRAY. You may question Mr. Lansdale as a witness, Mr. Garrison.

GARRISON. Was it up to you to grant Dr. Oppenheimer his clearance?

LANSDALE. Yes, or to refuse it. A difficult decision.

GARRISON. Why?

LANSDALE. In the opinion of the experts, Oppenheimer was the only person who could make Los Alamos a reality. But, on the other hand, the F.B.I. reports on him didn't look too good. The F.B.I. recommended that Dr. Oppenheimer should be withdrawn from the project. Thus I had to arrive at a judgment of my own.

GARRISON. How did you set about it?

LANSDALE. I had him placed under surveillance.

GARRISON. How was that done?

LANSDALE. We shadowed him, we opened his mail, we had his phone calls monitored, we set him various traps—well, we did all the nasty things that are usually done in such cases. And during that whole period I talked with him, and his wife, as often as I possibly could. I think he rather liked me. At any rate, he talked to me very frankly.

GARRISON. What was the purpose of these conversations?

LANSDALE. I wanted to find out what kind of a person he was, *what* he thought, and *how* he thought. I had to reach my own conclusion whether he was a Communist, as the F.B.I. suspected, or whether he was not.

GARRISON. What was your conclusion?

LANSDALE. That he was not a Communist, and that he should be granted his clearance no matter what the F.B.I. reports said.

GARRISON. Dr. Oppenheimer has been rebuked here for having refused to disclose the identity of his friend, Chevalier. What is your attitude in this matter?

LANSDALE. I thought it wrong and also rather naïve of him to imagine that he would get away with it, when he had *us* to deal with. His motives were that he regarded Chevalier as innocent and therefore wanted to protect him from trouble. Curiously enough, I always thought it was his brother, Frank, whom he wanted to protect, and General Groves thought the same.

GARRISON. Did his refusal endanger the security of the project?

LANSDALE. No. It just gave us a lot of extra work, that story he dished up to us. It was typical.

GARRISON. Typical of what?

LANSDALE. Scientists regard security officers either as extraordinarily stupid, or as extraordinarily cunning. And incompetent, in either case.

EVANS. Oh, how would you explain that?

LANSDALE. The scientific mind and the military security mind —well, it's like birds and rhinos sharing a ball game. Each thinks the other impossible, and both are quite right.

EVANS. Which are the rhinos?

LANSDALE. They are very nice animals.

GARRISON. Major Radzi has testified here that he still believes the true story to be the one about the three contacts, the microfilms, and the man at the Soviet Consulate.

LANSDALE. I know, but his views don't tally with the findings of our investigations.

GARRISON. Were those investigations concluded?

LANSDALE. They were concluded three times—in 1943, 1946, and 1950. I hope they will be concluded now, for the fourth time. The whole thing was just a lot of hot air.

GARRISON. If you had to decide now, today, whether to grant Dr. Oppenheimer his clearance, would you do so?

LANSDALE. By the same criteria which we used then, yes, certainly. I am not attempting to interpret the present ruling. Our criteria were: loyalty and discretion.

GARRISON. Thank you, Mr. Lansdale.

GRAY. Mr. Robb, do you wish to cross-examine Mr. Lansdale?

ROBB. Mr. Rolander will do so.

ROLANDER. As I understand it, sir, you are not offering any opinion as to whether or not you would clear Dr. Oppenheimer on the basis of presently existing criteria?

LANSDALE. These criteria are strange to me. I know them, but they appear strange to me. I do not wish to discuss their usefulness. On the basis of my past experiences with Dr. Oppenheimer, I regard him as perfectly loyal and very discreet.

ROLANDER. His discretion was good?

LANSDALE. Very good.

ROLANDER. Does your idea of discretion include spending the night with a Communist woman?

LANSDALE. Mr. Rolander, if you should ever fall deeply in love with a girl who happens to have Communist views, and if she wants to see you because she is unhappy—then, I hope, you will go and comfort her, and leave your tape recorder behind.

ROLANDER. You have not answered my question, sir.

LANSDALE. The question concerning Jean Tatlock has been answered seventeen times, Mr. Rolander! Dr. Oppenheimer was under our surveillance. I have heard the tape recordings and I have destroyed the tapes.

ROLANDER. Why?

LANSDALE. Because there is a limit to everything, Mr. Rolander!

ROLANDER. I fail to understand you, sir.

LANSDALE. That's just too bad.

GRAY. I suggest we regard this whole matter as concluded, Mr. Rolander.

ROBB. Does the name Steve Nelson mean anything to you?

LANSDALE. Yes.

ROBB. Who was he?

LANSDALE. A Communist functionary from California of whom it was said that, towards the end of 1943 or thereabouts, he had discovered that we were working on atomic weapons.

ROBB. Through whom was he supposed to have made that discovery?

LANSDALE. It was said that the F.B.I. suspected he might have got that information through Jean Tatlock or Mrs. Oppenheimer. Our investigations——

ROBB. Would you confine yourself to my question, Mr. Lansdale.

LANSDALE. May I finish my sentence? Our investigations did not uncover any evidence to support that conjecture.

ROBB. On the basis of your investigations, could such a possibility be entirely excluded?

LANSDALE. We found nothing to go on.

ROBB. But you would not commit yourself to saying that such a possibility can be entirely excluded?

LANSDALE. Have it your own way.

ROBB. I would like to ask Dr. Oppenheimer a question.

GRAY. Very well.

ROBB. Would you call Steve Nelson an intimate acquaintance of yours?

OPPENHEIMER. No. He was an acquaintance of my wife's. He had been in Spain with her first husband. He came to see us two or three times when he was at Berkeley, till about 1942.

ROBB. What did you talk about when he was there?

OPPENHEIMER. No idea. About personal matters. He came with his wife, I think.

ROBB. Did Jean Tatlock know him well?

OPPENHEIMER. Superficially. There was nothing personal in their relationship.

ROBB. In that case, if Jean Tatlock had in fact visited him, Doctor, the assumption would be that the reasons were purely political, is that right?

OPPENHEIMER. I cannot answer that. "Would"! "Could"! "Should"!

ROBB. Admittedly, Doctor, it is a hypothetical question. I shall now put it this way: If, let us suppose, Jean Tatlock had, through somebody, discovered something about our atomic weapons project—we are only supposing it, of course —then, with your knowledge of her psychological make-up, would you entirely exclude the possibility that she might have confided this secret to Steve Nelson?

OPPENHEIMER. She did not discover anything through me.

ROBB. Would you see no connection whatever between such a hypothetical visit and her tragic end? [OPPENHEIMER *does not answer.*] I am asking you a question, Doctor.

OPPENHEIMER. I know. And I am not answering it.

ROBB. Mr. Chairman . . .

GARRISON. Mr. Chairman . . .

GRAY. On the basis of a previous objection by the defense, which was sustained, the question put by Mr. Robb is inadmissible. Mr. Lansdale is on the stand as a witness.

ROLANDER. You have said here, sir, that in your opinion the story dished up by Dr. Oppenheimer was typical.

LANSDALE. His attitude was typical.

ROLANDER. Typical of what?

LANSDALE. Of scientists.

ROLANDER. Dr. Oppenheimer has testified here that he lied to you and to Major Radzi. Is telling lies characteristic of scientists?

LANSDALE. It is characteristic of them that they want to decide for themselves what information I need and what I don't.

ROLANDER. My question was whether you think that scientists, as a group, are liars.

LANSDALE. I don't think people as a group are liars. But it is the tendency of brilliant people to regard themselves competent in matters in which they have no competence.

ROLANDER. At the time, sir, it was a question of investigating what you believed to be a serious case of suspected espionage. Is that right?

LANSDALE. Yes. Well, yes.

ROLANDER. And Dr. Oppenheimer knew this when he refused to disclose Chevalier's name to you?

LANSDALE. Yes.

ROLANDER. And you told him that his refusal seriously impeded your investigation?

LANSDALE. He was neither the first nor the last scientist who impeded my investigations.

ROBB. Mr. Lansdale, do you feel you have to defend Dr. Oppenheimer here?

LANSDALE. I am trying to be as objective as possible.

ROBB. Your last answer made me doubt it.

LANSDALE [*losing his self-control*]. The questions put by this young man make me doubt whether it is the truth that is to be discovered here! I am extremely disturbed by the current hysteria, of which these questions are a manifestation!

ROBB. You think these proceedings are a manifestation of hysteria?

LANSDALE. I think——

ROBB. Yes or no?

LANSDALE. I refuse to answer with "yes" or "no"! If you are going to continue in this way . . .

ROBB. What?

LANSDALE [*regaining his self-control*]. If you will allow me to finish, I shall gladly answer your question.

ROBB. Very well.

LANSDALE. I think that the current hysteria over Communism is a danger to our way of life and to our form of democracy. Lawful criteria are being obliterated by fear and demagoguery. What is being done today, what so many people are doing today, is looking at events which took place in 1941, 1942, and judging them in the light of their present feelings. But human behavior varies in the changing context of time. If associations in the thirties or forties are regarded in the same light as similar associations would be regarded today—then, in my view, it is a manifestation of the current hysteria.

ROBB. It is true, then, that you regard these proceedings as a manifestation of——

LANSDALE. Hell, I was told off, at the time, because I refused to allow the former political officer of the Lincoln Brigade to serve in our army—and then he was commissioned on direct orders from the White House! That's how it was then. What is the use of rehashing old stuff dating back to 1940 or 1943? That is what I mean by hysteria.

ROBB. How do you know that this Board is rehashing old stuff?

LANSDALE. I don't know. I hope I am wrong.

ROLANDER. Would it be true to say, sir, that the security officers below your own rank unanimously opposed the clearance of Dr. Oppenheimer?

LANSDALE. If I had judged by the F.B.I. reports alone, I would also have opposed it. But the success of Los Alamos, the atomic bomb—that was Dr. Oppenheimer.

ROLANDER. Thank you, sir.

GRAY. Any further questions to the witness? Mr. Morgan.

MORGAN. Mr. Lansdale, when you reached the conclusion that Dr. Oppenheimer was not a Communist, what was your definition of a Communist?

LANSDALE. A person who is more loyal to Russia than to his own country. You will note that this definition has nothing to do with philosophical or political ideas.

MORGAN. What was the alignment of Dr. Oppenheimer's political ideas?

LANSDALE. They were extremely liberal.

MORGAN. Do you think this can always be differentiated from Red itself?

LANSDALE. In many cases, no.

MORGAN. If I understood you correctly: contrary to Major Radzi, you do not think that Dr. Oppenheimer has discredited himself by his behavior in the Chevalier incident?

LANSDALE. No.

MORGAN. I am an old businessman, a pragmatist. May I ask you a hypothetical question?

LANSDALE. Please do.

MORGAN. Let us assume you are the president of a large bank.

LANSDALE. Gladly!

MORGAN. Would you employ a man at your bank who is on intimate terms with safe-crackers? Would you employ him as the manager of your bank?

LANSDALE. If he is first-class.

MORGAN. Well, let us say you have such an excellent bank manager. . . . One day, a friend of this bank manager comes to see him, and he says: "I know some efficient fellows who'd be very interested in robbing your bank. Nothing can go wrong. All you have to do is disconnect the alarm system for a while." Your bank manager rejects the proposition; let us assume he uses strong words. Supposing he does not report this incident to you for six months— and he reports it only in connection with, let us say, some bank robbery in Chicago—would this not surprise you?

LANSDALE. I would ask him why he had waited so long before telling me.

MORGAN. Supposing he says to you: "The man who approached me about it is a good friend of mine; I didn't take it seriously; I'm sure he himself has nothing to do with it; that's why I didn't want to get him into trouble. But, because of that Chicago business, I want to draw your attention to the guy who took the initiative in the whole thing, at the time." Would you not urge him to tell you also the name of his friend?

LANSDALE. I guess I would. Of course I'd also try to discover whether it was a serious matter or idle talk.

MORGAN. Now let us suppose he tells you the story in this way: "My friend told me at the time that the fellows he knows intend to break into quite a number of banks, using all the latest gadgets." Would you not have reached the conclusion that the police ought to be informed of this?

LANSDALE. Yes.

MORGAN. Well, your bank manager is now put under pressure to disclose the name of his friend. He comes to you and says: "Mr. Lansdale, I recently told you a story about my friend and those other fellows. Tear gas, submachine guns, and so forth. Well, it was just a cock-and-bull story. Nothing in it is true. I merely wanted to protect my friend from trouble." Would you not ask yourself: why? What is behind it? Does one protect one's friend by telling an extraordinary cock-and-bull story about him?

LANSDALE. I certainly would have asked myself these questions. But I wouldn't have done so twelve years later, when it turned out that none of those fellows had actually robbed a bank.

MORGAN. Do you know all the banks in America, Mr. Lansdale?

LANSDALE. The one you refer to is a bank I know quite well. The analogy does not fit.

MORGAN. I agree it is crude. Crude thinking is one of my most profitable faculties.

GRAY. Any further questions to Mr. Lansdale? Yes, Mr. Evans.

EVANS. I already asked Mr. Radzi. I wasn't satisfied with his answer; perhaps the fault lies with the question itself: With reference to a secret war project, is there such a thing as one-hundred-per-cent security?

LANSDALE. No.

EVANS. Why is that so?

LANSDALE. In order to have one-hundred-per-cent security, we would have to abandon all the freedoms we want to defend. It can't be done.

EVANS. What, in your opinion, can be done, then, to ensure a country the maximum amount of security?

LANSDALE. We must see to it that we have the best ideas and the best way of life.

EVANS. I'm no expert, but I feel I'd also have formulated it along these lines. . . . It isn't easy.

LANSDALE. No.

EVANS. That is all.

GRAY. Thank you very much, Mr. Lansdale.

LANSDALE *rises to his feet.*

EVANS. One more question perhaps, just as nonprofessional or naïve as my last one. When I look at the results of this strict secrecy, of these security ramifications on all sides, I mean, we are sitting rather uncomfortably on this powder keg of a world, everywhere, I think the question arises whether these secrets might not be safeguarded best of all by being made public?

LANSDALE. What do you mean?

EVANS. By re-establishing the age-old right of the scientists to publish the results of their researches, or maybe even demanding it of them?

LANSDALE. This is such a remote and utopian dream, Dr. Evans, that even children are forbidden it. . . . The world is divided into sheep and goats, and we are all in the slaughterhouse.

EVANS. As I said before, I'm no expert.

GRAY. Thank you very much, Mr. Lansdale. [LANSDALE *leaves the room.*] Today's session is concluded, and we will now take a recess. We shall next deal with Dr. Oppenheimer's attitude in the matter of the hydrogen bomb. May I ask Mr. Garrison and Mr. Robb for the list of their witnesses.

PART TWO

The stage is open, as before. The following documentaries are projected on the hangings, with a simultaneously spoken text:

PROJECTION	SPOKEN TEXT
October 31, 1952. Test explosion of the first hydrogen bomb in the Pacific.	Test explosion of Mike, the first hydrogen bomb, in the Pacific.
The island of Elugelab sinks into the ocean.	The island of Elugelab, Atoll Eniwetok, sinks into the ocean.
President Truman speaks. Applause from a large crowd.	President Truman announces the American monopoly of the hydrogen bomb.
August 8, 1953. Test explosion of the first Russian hydrogen bomb.	Test explosion of the first Russian hydrogen bomb in Soviet Asia.
Minister President Malenkov speaks. Applause from a large crowd.	Minister President Malenkov declares: "The United States no longer holds the monopoly of the hydrogen bomb."
An American fleet of bombers. A Soviet fleet of bombers.	In the present state of nuclear balance, the high commands of the two big world powers keep their strategic A- and H-bomber fleets in the air.

The hangings close.

SCENE 1

The members of the Board and counsel for both parties occupy their usual places. OPPENHEIMER *is on the witness stand. The Chairman,* GORDON GRAY, *steps forward to the footlights.*

GRAY. I've been afraid it might happen, and now it has. The *New York Times* has published the letter of the Atomic Energy Commission and Oppenheimer's reply to the allegations. The letters were released by Oppenheimer's counsel in order to counteract a subversive campaign against Oppenheimer. Now the case of J. Robert Oppenheimer dominates the headlines and public discussion all over America.

With a resigned gesture, he returns to his seat.

Voices reciting the following headlines issue from a loudspeaker. Simultaneous with the voices, five photographs of OPPENHEIMER *are successively projected, very different from each other, the facial expression in each corresponding with the headline that goes with it.*

VOICES ISSUING FROM THE LOUDSPEAKER:

The man who rated his personal friendships higher than loyalty to his country. [*Corresponding photo.*]

The man who betrayed his friends because of his loyalty to his country. [*Corresponding photo.*]

The martyr who fought against the development of the hydrogen bomb on moral grounds. [*Corresponding photo.*]

The ideological traitor who destroyed America's nuclear monopoly. [*Corresponding photo.*]

Oppenheimer, an American Dreyfus Case. [*Corresponding photo.*]

End of projection of photographs.

SCENE 2

The following text is projected on the hangings:

THE PROCEEDINGS ENTER A DECISIVE STAGE.

LOYALTY TO A GOVERNMENT.

LOYALTY TO MANKIND.

ROBB. We would now like to deal with your attitude in the matter of the hydrogen bomb, Doctor.

OPPENHEIMER. All right.

ROBB. I quote from the letter by the Atomic Energy Commission, page six, at the bottom: "It was further reported that in the fall of 1949, and subsequently, you strongly opposed the development of the hydrogen bomb; (1) on moral grounds, (2) by claiming that it was not feasible, (3) by claiming that there were insufficient facilities and scientific personnel to carry on the development, and (4) that it was not politically desirable." Is this statement true?

OPPENHEIMER. Partly. With reference to a specific situation in the fall of 1949, and to a specific technical program.

ROBB. Which parts of it are true, Doctor, and which are not?

OPPENHEIMER. I made that clear in my answering letter.

ROBB. I would like to have it clearer still.

OPPENHEIMER. Let us try.

ROBB. I have a report here, from the General Advisory Committee of which you were the Chairman. It dates back to October 1949, and it is in answer to the question whether the United States should, or should not, initiate a crash program for the development of the hydrogen bomb. Do you remember that report? [*He hands* OPPENHEIMER *a copy.*]

OPPENHEIMER. The majority report. I wrote it myself.

ROBB. It says there—Mr. Rolander will read it out to us . . .

ROLANDER. "The fact that there are no limits to the destructiveness of this weapon makes its very existence a danger to humanity as a whole. For ethical reasons, we think it wrong to initiate the development of such a weapon."

OPPENHEIMER. That is from the minority report which was written by Fermi and Rabi.

ROLANDER. It says here, in the majority report: "We all hope that the development of this weapon can be avoided. We are all agreed that it would be wrong at the present moment for the United States to initiate an all-out effort toward the development of this weapon."

ROBB. Does this not mean, Doctor, that you were against the development of the hydrogen bomb?

OPPENHEIMER. We were against *initiating* its development. In an exceptional situation.

ROBB. What was exceptional about the situation in the fall of 1949, Doctor?

OPPENHEIMER. The Russians had exploded their first atomic bomb, Joe I, and we reacted with a nation-wide shock. We had lost our monopoly of the atomic bomb, and our first reaction was: we must get a hydrogen bomb monopoly as quickly as possible.

ROBB. That was quite a natural reaction, was it not?

OPPENHEIMER. Maybe natural, but not sensible. The Russians then also developed the hydrogen bomb.

ROBB. Were we not in a much better position, technically?

OPPENHEIMER. Perhaps, but in Russia there are only two targets suitable for a hydrogen bomb, Moscow and Leningrad, whereas we have more than fifty.

ROBB. One more reason to get ahead of the Russians, or is it not?

OPPENHEIMER. It seemed wiser to me to try for an international declaration of renunciation of that terrible weapon. With a third world war, fought with hydrogen bombs, there would be no victors and no vanquished any more but only the destruction of ninety-eight per cent or a hundred per cent of mankind.

MORGAN. A declaration of renunciation without controls? I think, Dr. Oppenheimer, you were the Scientific Adviser of our government when Gromyko declared in Geneva, in 1946, that he could not agree to any kind of control whatsoever. And, in those days, we had the monopoly of the atom bomb.

OPPENHEIMER. Yes, I felt very depressed at the time.

MORGAN. Why should the Russians be more obliging in 1949?

OPPENHEIMER. The ability to extinguish all life upon earth is a new attribute. The writing on the wall has appeared for mankind.

MORGAN. In Cyrillic lettering, too, Doctor?

OPPENHEIMER. Since we had the fall-out of the Russian hydrogen bomb analyzed, yes. We should have knocked before opening the door to the horrible world in which we now live. But we preferred to break the door down. Although it gave us no strategic advantage.

MORGAN. Did you feel competent to decide strategic questions? Was that your function?

OPPENHEIMER. The greater part of our report gave our assessment of whether a serviceable hydrogen bomb could be made, and how long it would take.

ROBB. How did you assess it?

OPPENHEIMER. We doubted the practicability of the technical propositions we had at that time—they proved impracticable indeed.

ROBB. Did this not mean that the hydrogen bomb was to be shelved until some better ideas emerged?

OPPENHEIMER. No. We recommended a research program.

ROBB. In other words, the project of the hydrogen bomb was in a bad way?

OPPENHEIMER. The first model looked pretty bad. Miserable. Otherwise we wouldn't have spoken of five years' development.

ROBB. Was the prognosis correct?

OPPENHEIMER. For that model?

ROBB. For the hydrogen bomb?

OPPENHEIMER. No. There were some brilliant ideas in 1951, and we tested Mike, the first hydrogen bomb, in October 1952.

ROBB. The test was very successful, was it not?

OPPENHEIMER. Yes. The island of Elugelab in the Pacific disappeared within ten minutes. Nine months later, the Russians had their hydrogen bomb. It was superior to our own model.

EVANS. In what respect, Dr. Oppenheimer?

OPPENHEIMER. The Russians had exploded the so-called "dry" hydrogen bomb which weighed much less than ours because it needed no cooling devices.

EVANS. Was that very essential, strategically?

OPPENHEIMER. Sure. The Russians could appear above us with their hydrogen bombs at any time, while we could only retaliate with atom bombs. Our first models were so heavy that we could get them to a target only by ox cart.

ROBB. Wouldn't the Russians have produced their hydrogen bomb in any case?

OPPENHEIMER. Possibly. We did not attempt to prevent an armaments race in that sphere. I think we paid too high a price for our short-lived monopoly.

ROBB. Couldn't we have had the hydrogen bomb much earlier, and wouldn't our position have been very different, if we had initiated the crash program in 1945?

OPPENHEIMER. We didn't have the facilities.

ROBB. In 1942, you already foresaw the possibility of developing a thermonuclear bomb, is that right?

OPPENHEIMER. We would have developed it if we could. We would have developed any kind of weapon.

ROBB. I do not know whether this is classified or not—but when we speak of a hydrogen bomb we mean a bomb with ten thousand times the power of the atomic bomb, is that right?

OPPENHEIMER. Roughly speaking, yes. Very powerful anyway.

ROBB. It would be no exaggeration to say: ten thousand times as powerful?

OPPENHEIMER. I believe there are no limits to its power. According to our calculations, the lethal zone of a medium-sized model has a diameter of about 360 miles.

ROBB. When you first went to Los Alamos, would you have had moral scruples about developing such a weapon?

OPPENHEIMER. In 1942? No. My scruples came much later.

ROBB. When? When did you start having moral scruples about developing the hydrogen bomb?

OPPENHEIMER. Let us leave the word "moral" out of it.

ROBB. All right. When did you start having scruples?

OPPENHEIMER. When it became clear to me that we would tend to *use* the weapon we were developing.

ROBB. After Hiroshima?

OPPENHEIMER. Yes.

ROBB. You have testified here that you helped to select the targets in Japan. Is that right?

OPPENHEIMER. Yes. And I said it had not been *our* decision that the bomb should be dropped.

ROBB. I am not saying that it was. You only selected the targets, and you had strong scruples after the dropping of the atomic bomb?

OPPENHEIMER. Yes. Terrible ones. We all had terrible scruples.

ROBB. Wasn't it those terrible scruples, Doctor, which prevented you from initiating an all-out hydrogen bomb program in 1945?

OPPENHEIMER. No. When the production of the hydrogen bomb seemed feasible in 1951 we were fascinated by the scientific ideas, and we produced it in a short time, regardless of scruples. That is a fact. I'm not saying it is a good fact.

ROBB. Did you do any scientific work on the hydrogen bomb?

OPPENHEIMER. Not any practical work.

ROBB. In what way did you contribute?

OPPENHEIMER. In an advisory capacity.

ROBB. Would you give us an example?

OPPENHEIMER. In 1951, I called a conference of the leading physicists. It was very fruitful. We were enthusiastic about the new possibilities, and many physicists returned to Los Alamos.

ROBB. Who was it who had the brilliant ideas?

OPPENHEIMER. Teller, mainly. Neumann's calculating machines played a part in it. Bethe and Fermi contributed.

ROBB. Did *you* return to Los Alamos?

OPPENHEIMER. No.

ROBB. Why not?

OPPENHEIMER. I had other work to do. . . . My scientific contributions in the thermonuclear field were negligible.

ROLANDER [*extracts a document from his files*]. I have a patent here, sir, for an invention appertaining to the hydrogen bomb. You applied for it in 1944.

OPPENHEIMER. Together with Teller?

ROLANDER. Yes. You were granted the patent in 1946.

OPPENHEIMER. That's right. It was a detail. . . . I had forgotten we followed the matter up.

ROBB. You refused Teller's request to come to Los Alamos, saying you wished to remain neutral in regard to the hydrogen bomb?

OPPENHEIMER. It is possible.

ROBB. That you wished to remain neutral?

OPPENHEIMER. That I said something to that effect. There was a time when Teller wanted to pursue the hydrogen bomb program at any price. I had to consider the arguments for and against—at least until the President himself ordered the crash program.

ROBB. But you refused to return to Los Alamos, even after the President's decision?

OPPENHEIMER. Yes.

ROBB. Don't you think, Doctor, that it would have made a

great impression on many scientists if you had rolled up your sleeves and had taken charge of the hydrogen bomb program yourself?

OPPENHEIMER. Maybe. I did not think it right.

ROBB. You did not think it right to produce the hydrogen bomb, even after the President's decision?

OPPENHEIMER. I did not think it right to take the responsibility for the program. I was not the right man for the job.

ROBB. That is not what I asked you, Doctor.

OPPENHEIMER. I feel you did.

ROBB. You did not think it right to produce the hydrogen bomb, even after the President's decision?

OPPENHEIMER. I always regarded it as a dreadful weapon, and that it would be better if it did not exist. But I supported the crash program.

ROBB. In what way?

OPPENHEIMER. In an advisory capacity.

ROBB. Any other way?

OPPENHEIMER. I recommended a number of young scientists, my pupils, to Teller.

ROBB. Did you talk to them? Did you make them feel enthusiastic about the program?

OPPENHEIMER. Teller talked to them, I don't know whether he made them enthusiastic.

ROBB. Didn't you say, Doctor, that you were enthusiastic about the program in 1951?

OPPENHEIMER. I was enthusiastic about the fascinating scientific ideas.

ROBB. You thought the scientific ideas for the development of the hydrogen bomb were fascinating and wonderful—and you thought of the possible result, the hydrogen bomb itself, as horrible. Is that right?

OPPENHEIMER. I think that's right. It isn't the fault of the physicists that brilliant ideas always lead to bombs nowadays. As long as that is the case, one can have a scientific enthusiasm for a thing and, at the same time, as a human being, one can regard it with horror.

ROBB. I see you are capable of it, Doctor. It surprises me.

GRAY. Would you not say, Dr. Oppenheimer, that such an attitude might imply something like divided loyalty?

OPPENHEIMER. Divided between whom?

GRAY. Loyalty to a government—loyalty to mankind.

OPPENHEIMER. Let me think. . . . I would like to put it this way: if governments show themselves unequal to, or not sufficiently equal to, the new scientific discoveries—then the scientist *is* faced with these conflicting loyalties.

GRAY. If you are facing such a conflict, Dr. Oppenheimer, and it obviously did happen in the case of the hydrogen bomb, to which loyalty would you give the preference?

OPPENHEIMER. In every case, I have given undivided loyalty to my government, without losing my uneasiness or losing my scruples, and without wanting to say that this was right.

ROBB. You do not want to say that it is right to give one's undivided loyalty to the government in every case?

OPPENHEIMER. I don't know. I think about it. But I have always done it.

ROBB. Does this apply also to the hydrogen bomb program?

OPPENHEIMER. Yes.

ROBB. You mean to say you actively supported it after the President's decision?

OPPENHEIMER. Yes, although my grave doubts remained.

ROBB *extracts a document from his files.*

ROBB. But you say in a television interview—Mr. Rolander will read it out to us . . .

He gives ROLANDER *the sheet of paper and places a copy on the Chairman's table.*

ROLANDER. I quote: "Early history tells us about the annihilation of individual tribes, individual races, individual nations. Now mankind as a whole can be annihilated by man. To the rational mind, this is now a probability unless we develop the new forms of political co-existence which are needed on this earth. That apocalyptic possibility is very real in our lives. We know it, but we suppress our knowledge. It does not seem immediate to us. We imagine there is still time. But we do not have much time left."

ROBB. Does that sound as though you supported the hydrogen bomb program, Doctor?

OPPENHEIMER. It had nothing to do with that program. When I gave the interview we had already tested our first models, and so had the Russians.

ROLANDER. That is not so, sir. You gave the interview before the presidential election of 1952, when we actually had the monopoly.

ROBB. I think this makes a difference. It was at the time, was it not, when the war in Korea had ended and when our bases in Asia seemed extremely vulnerable?

OPPENHEIMER. It was at a time when quite a number of people were seriously discussing the idea of a preventive war, yes.

ROBB. Did *you* discuss that idea?

OPPENHEIMER. Well, we were asked to give our opinion on a technical problem, and we reached a negative conclusion.

MORGAN. A question of conscience, Dr. Oppenheimer: if it had looked pretty good from a technical point of view, would you have confined yourself to a technical opinion only?

OPPENHEIMER. I don't know—I hope not. No.

ROBB. Doesn't it emerge quite clearly from our conversation here, Doctor, that you had strong moral scruples about the hydrogen bomb, and that you still have them today?

OPPENHEIMER. I have already asked you to leave the word "moral" out of it. It confuses the issue. I had, and I have, grave scruples about this dreadful weapon actually being used.

ROBB. And that is why you were against the development of the hydrogen bomb. Is that right?

OPPENHEIMER. I was opposed to taking the initiative in its development.

ROBB. Doctor, in the report of the General Advisory Committee, which you yourself wrote—and in the appendix, which you approved—did you not state quite clearly, I quote: [*He reads it out.*] "We believe a hydrogen bomb should never be produced."

OPPENHEIMER. That referred to the program at the time.

ROBB. What did you mean by "never"?

OPPENHEIMER. I did not write the appendix.

ROBB. But you signed it, did you not?

OPPENHEIMER. I believe what we meant to say, what I meant to say, was that it would be a better world if there was no hydrogen bomb in it.

ROBB. What was your reaction when, in spite of it, the President ordered the crash program?

OPPENHEIMER. I offered to resign from the General Advisory Committee.

ROBB. As a protest?

OPPENHEIMER. I guess a man has to take the consequences when he has been overruled on a crucial issue.

ROBB. You felt you had been overruled when the order was given to initiate the crash program?

OPPENHEIMER. We had advised against it. Yes.

ROBB. When the hydrogen bomb was to be tested in October 1952, would it be true to say that you opposed that test?

OPPENHEIMER. "Opposed" is too strong. I was in favor of postponement.

ROBB. Why?

OPPENHEIMER. It was shortly before the presidential election, and I did not think it right to involve the new President with the responsibility for the hydrogen bomb. He should make his own decision.

ROBB. Were there any other reasons in favor of postponement?

OPPENHEIMER. The Russians would have got a lot of information out of the test.

ROBB. Any other reasons?

OPPENHEIMER. It would most certainly have buried our hopes for disarmament negotiations, particularly our hopes for a test ban.

ROBB. But, in spite of your recommendations, the hydrogen bomb was tested in October 1952?

OPPENHEIMER. Yes.

ROBB. If, in the style of our newspaper columnists, we were to speak of the Father of the Hydrogen Bomb—whom would you call that?

OPPENHEIMER. Teller has been called that.

A photograph of TELLER *is projected.*

ROBB. You yourself would lay no claim to that title?

OPPENHEIMER. Oh, no.

ROBB. Thank you, Dr. Oppenheimer.

GRAY. Any further questions to Dr. Oppenheimer? . . . Mr. Morgan.

MORGAN. I have only one question, Dr. Oppenheimer. When a country spends enormous sums of money on research, is it to be denied the right to do with the results of that research as it thinks fit?

OPPENHEIMER. Since certain results of research can be used to destroy human civilization as a whole, such a right has become questionable.

MORGAN. Does this not mean that you want to limit the national sovereignty of the United States in this particular sphere?

OPPENHEIMER. When things have gone so far that mathematicians have to calculate whether a certain test might not burn up the whole atmosphere—well, then national sovereignties begin to look slightly ridiculous. The question is, what authority is independent and powerful enough to prevent nations, or groups of nations, from committing suicide? How is such an authority to be established?

MORGAN. Do you think the United States should make an effort towards reaching some understanding with Soviet Russia?

OPPENHEIMER. If the Devil himself were on the other side, one would have to reach an understanding with the Devil.

MORGAN. But you do draw a sharp distinction between the preservation of life as such and the preservation of a life that is worth living?

OPPENHEIMER. Oh, yes. And I have great faith in the ultimate power of common sense.

EVANS. I would like to refer back to the moral scruples, so often mentioned here, and to the contradiction in forging ahead with something and, at the same time, being afraid of the result. When did you first feel this contradiction?

OPPENHEIMER. When we exploded the first atom bomb in the Alamogordo desert.

EVANS. Would you elucidate that?

OPPENHEIMER. When I saw that ball of fire, two passages in the Bhagavad-Gita came to my mind. One was:

> "The radiance of a thousand suns
> which suddenly illuminate the heavens
> all in one moment—thus
> the splendor of the Lord."

And the other:

> "And I am Death, who taketh all,
> who shatters worlds. . . ."

EVANS. How can you tell whether a new idea is really important?

OPPENHEIMER. I know it from the profound awe I feel.

GRAY. If there are no further questions I would like to thank Dr. Oppenheimer for his patience. [OPPENHEIMER *leaves the witness stand and walks across to the sofa.*] We now have the witnesses who have been called by Mr. Robb and Mr. Garrison. Dr. Teller has already been waiting some time. Therefore, I think, we should hear him first, and Mr. Griggs afterwards.

An OFFICIAL *leaves the room to fetch* TELLER.

GARRISON. If possible, Mr. Chairman, we would prefer to call Dr. Bethe as a witness after Dr. Teller.

GRAY. Indeed it is possible. Can you get in touch with Dr. Bethe?

GARRISON. He is waiting in his hotel. It wouldn't take more than five minutes. Here is his telephone number.

He hands another OFFICIAL *a slip of paper, and the* OFFICIAL *leaves the room. The first* OFFICIAL, *returning, appears at the door with* TELLER.

GRAY. If Dr. Teller is ready, may we ask him to take the stand as a witness. Dr. Teller, do you wish to testify under oath?

TELLER. Yes.

TELLER *rises to his feet.*

GRAY. Edward Teller, do you swear that the testimony you are to give the Board shall be the truth, the whole truth, and nothing but the truth, so help you God?

TELLER. I do.

GRAY. Mr. Robb, you may question the witness.

ROBB. Dr. Teller, it was already at Los Alamos that you were working on the problem of thermonuclear development?

TELLER. Yes.

ROBB. Did you frequently discuss the thermonuclear problems with Dr. Oppenheimer?

TELLER. Very frequently. Ever since we met at Berkeley, in the summer of 1942. We investigated the feasibility of a thermonuclear program.

ROBB. Whom do you mean by "we"?

TELLER. Oppenheimer had gathered together the best people in that field. Among them, Fermi and Bethe. The idea of reproducing the miracle of solar energy by a fusion of light-weight atoms filled us all with enthusiasm.

ROBB. Was Dr. Oppenheimer enthusiastic?

TELLER. Very. And he has the ability to rouse others to enthusiasm.

ROBB. Did a thermonuclear program seem feasible to you then?

TELLER. For a time, it seemed easier than it actually turned out to be. We encountered serious difficulties at Los Alamos. I think I myself discovered some of them.

ROBB. Could you tell us about some of the difficulties, without violating any secrecy?

TELLER. One of them was that we needed the heat of an ordinary atomic bomb to get a hydrogen bomb going. Later we discovered how to generate the required heat in a different way. Another difficulty was that our conventional calculating machines proved inadequate. And so forth.

ROBB. In spite of all this, would it have been possible to produce the hydrogen bomb at Los Alamos during the war?

TELLER. No. I rather fancied the idea, it was my baby, but parents are shortsighted.

ROBB. In your opinion, when were there the facilities to initiate an all-out hydrogen bomb program?

TELLER. In 1945. I remember, after "Trinity," that we——

EVANS. What is that?

TELLER. "Trinity" was the code name for the atomic bomb test in Alamogordo.

EVANS. "Trinity"?

TELLER. Yes. . . . I remember that, afterwards, we wanted to develop the hydrogen bomb in a very vigorous fashion— the most outstanding people, like Fermi and Bethe, under Oppenheimer's direction.

ROBB. After the first atomic bomb test?

TELLER. Yes.

ROBB. Was work on the development in fact stepped up?

TELLER. No. In a very short time, it all came to a standstill.

ROBB. Why?

TELLER. After the dropping of the atom bombs on Japan, the whole plan was practically abandoned.

ROBB. Why?

TELLER. Because, after Hiroshima, Dr. Oppenheimer thought it was not the time to pursue such a program any further.

ROBB. Did he say that to you?

TELLER. I remember a conversation with Oppenheimer, Fermi——

ROBB. Was Fermi of the same opinion?

TELLER. Yes. . . . I should add, however, that this was in consonance with the general mood of the physicists. Hiroshima had been a great shock for many of them. The mood in those days was something like a hangover.

ROLANDER. Did Dr. Oppenheimer say that the best thing to do would be to give Los Alamos back to the Indians?

TELLER. He is said to have made that remark, I don't know whether he actually made it.

ROBB. Would it have been possible, at Los Alamos, to initiate a crash program for the hydrogen bomb by the end of the war?

TELLER. It is my firm belief that we would have been in a position to initiate an energetic hydrogen bomb program. If Dr. Oppenheimer had stayed on at Los Alamos, if he had supported that program, other highly capable people would also have participated—at least as many people as we had recruited in 1949 under much more difficult conditions.

ROBB. In that case, we would have had the hydrogen bomb sooner?

TELLER. I am convinced of that.

ROBB. When, according to your estimate, could we have had the hydrogen bomb?

TELLER. It is very difficult to assess when that would have been possible if things had taken a different course. To conjecture about the past is almost as difficult as to make predictions about the future, except that it is less risky to do so.

ROBB. Let us try, all the same.

TELLER. If we had initiated the program in 1945 we would probably have had the hydrogen bomb in 1948.

ROBB. Before the Russians had their atomic bomb?

TELLER. Presumably.

ROBB. It has been said here, Doctor, that the ultimate success of the program was the result of a brilliant discovery which you made in 1951. Is that so?

TELLER. If such excellent people as Fermi, Bethe, and others, had gone after the program in 1945 they'd probably have had the same brilliant idea, or other brilliant ideas. Then we'd have had the hydrogen bomb as early as 1947.

ROBB. You mean: "If one doesn't seek, one doesn't find." Is that right?

TELLER. Brilliant ideas can be organized. They are not tied up with particular individuals.

GRAY. What would have been the advantages of having the hydrogen bomb in, let us say, 1947?

TELLER. You will know that better than I do, from the vantage point of the War Department. It would have saved us the débâcle in China and, presumably, a few other disasters. We would have kept our number one position in relation to the Communists, and that is a comfortable position, I should think.

MARKS. Do you know, Dr. Teller, that our secret services found out that the Russians had reached roughly the same stage in research as we had, in 1945?

TELLER. Yes. That is why I wanted the hydrogen bomb, while other people were playing around with disarmament illusions.

MARKS. By other people you mean the government at that time, don't you?

TELLER. The government, the physicists, public opinion. It was enough to drive one to despair.

ROBB. When did you leave Los Alamos?

TELLER. In February 1946. There was no point in staying

there. I accepted a teaching post at the University of Chicago and went back only occasionally to Los Alamos, as a consultant.

ROBB. Would you tell us about the work being done on thermonuclear development at Los Alamos between 1945 and 1949?

TELLER. It was virtually at a standstill. It was resumed only in 1949 when the Russians had exploded their atom bomb.

ROBB. Did you then talk with Oppenheimer?

TELLER. Yes. Frankly, I was staggered.

ROBB. Why?

TELLER. In those days, I had so little to do with war projects that I only read in a newspaper about the Russian atom bomb. I decided that I must devote myself entirely to an effective hydrogen bomb program, no matter what the odds. I called up Dr. Oppenheimer, and I said: "For God's sake what's going to happen now?" And I asked for his advice. Do you know what he said? He said: "Keep your shirt on!"

ROBB. What did you conclude from that advice?

TELLER. That a hydrogen bomb program could only be undertaken against his wishes, and that it would be a big problem, considering Oppenheimer's influence.

ROLANDER. Did you discuss it further?

TELLER. Yes.

ROLANDER. On what occasion?

TELLER. I discussed things with Bethe. We had to get a team together, and I had strong hopes that Bethe would agree to take charge of the hydrogen bomb program.

ROLANDER. When was that?

TELLER. At the end of October.

ROBB. 1949?

TELLER. Yes. Shortly before the General Advisory Committee decided against the crash program. I insisted so much that Bethe finally made up his mind to come to Los Alamos, even though he had strong reservations, at least that was my impression. In the middle of all this, Oppenheimer called us up and invited us to Princeton. I said to Bethe: "After you've talked with Oppenheimer you won't come."

ROLANDER. Did Dr. Bethe come to Los Alamos?

TELLER. No. Only much later.

ROLANDER. Do you think it was due to Dr. Oppenheimer's influence that Dr. Bethe didn't come when you first asked him to?

TELLER. Yes. When we left Oppenheimer's office, Bethe said: "You can be quite satisfied, I'm still coming." Two days later he called me up, and he said: "Edward, I've thought it over. I'm not coming."

ROLANDER. Do you know whether Dr. Bethe had in the meantime again talked with Dr. Oppenheimer?

TELLER. I guess he had.

ROBB. Did Dr. Oppenheimer bring forward any moral or political arguments against the program?

TELLER. He quoted the arguments of other people, for and against; for instance, there was a letter from Conant which said: "The hydrogen bomb—only over my dead body."

ROBB. When the General Advisory Committee decided against the crash program, would you say that this decision was due to Dr. Oppenheimer?

TELLER. That would be going too far.

ROBB. Was the assessment of the technical side of the program correct?

TELLER. It was not correct in so far as it overlooked the great possibilities of development, which we were soon able to verify.

ROBB. Do you think it possible that the technical snags rather suited some of the members?

TELLER. Not consciously, I don't think.

ROBB. Unconsciously?

TELLER. That is too vague a question.

ROBB. What effect did the report have on the physicists who were working on the hydrogen bomb?

TELLER. Rather a paradoxical one. When I saw the report, and ten or twelve others were also shown it, on Oppenheimer's suggestion, I thought it was sure to be the end of the program. To my surprise, there was a psychological reaction of defiance among those who participated in the program.

ROBB. You mean the report infuriated them, and they were more than ever determined to continue their work?

TELLER. Yes, it made them indignant that their work was supposed to be immoral as soon as it was making good progress.

ROBB. Did Dr. Oppenheimer support the crash program when it had finally been ordered by the President?

TELLER. I can't remember any such support, on the contrary.

ROBB. When you say "on the contrary" you mean he continued to oppose the crash program?

TELLER. I mean that the further recommendations of the General Advisory Committee served as a brake rather than encouragement.

ROLANDER. Would you give us some examples?

TELLER. The second laboratory. We wanted to concentrate the program at Livermore; the General Advisory Committee was against it. We wanted to develop reactor work for our purposes at Oak Ridge; the General Advisory Committee concentrated it in Chicago. We needed more money because only the experimental approach could bring us nearer a solution; Oppenheimer recommended further theoretical research without tests. All this did not further our work, but impeded it.

ROBB. Did you have any discussions with Dr. Oppenheimer during that time?

TELLER. A few.

ROBB. How would you describe his attitude?

TELLER. Reserved and neutral. He told me that himself when

I asked him for advice about the recruitment of capable people.

ROBB. Did he give you the advice you asked for?

TELLER. I wrote to all the people he had suggested; none of them came. . . . But I should add that Dr. Oppenheimer's attitude to the program changed at a later stage.

ROBB. When was that?

TELLER. In 1951, after our first tests. Then he called a meeting of the General Advisory Committee and all the experts, in Princeton. I went to that meeting with considerable misgivings because I expected some further obstacles. But Oppenheimer was actually delighted with our new theoretical discoveries, and he said he'd never have opposed the program if those wonderful ideas had come earlier.

ROBB. Did he then support the program?

TELLER. Not to my knowledge, but he may of course have assisted without my noticing it.

ROBB. A question to you as an expert: if Dr. Oppenheimer were to go fishing for the rest of his life, what effect would that have on the further atomic energy program?

TELLER. Do you mean if he did similar work to what he did at Los Alamos, or similar to what he did after the war?

ROBB. Similar to what he did after the war.

TELLER. After the war, Dr. Oppenheimer mainly served on committees and, judging from my own experience, I would say that all committees can go fishing without affecting the work of those who are actually doing the work.

ROBB. These are my questions. Thank you for having given us your valuable time.

GRAY. Do you wish to cross-examine the witness, Mr. Garrison?

GARRISON. Mr. Marks has some questions.

MARKS. Dr. Teller, in your opinion, has Dr. Oppenheimer been disloyal to the United States?

TELLER. Until I am given proof to the contrary, I shall believe that he wanted to act in the best interests of the United States.

MARKS. You regard him as perfectly loyal?

TELLER. Subjectively, yes.

MARKS. Objectively?

TELLER. Objectively, he has given wrong advice which was injurious to this country.

MARKS. Is the loyalty of a man, who gave great service, to be doubted because he came forward with advice which was later regarded as having been wrong?

TELLER. No, but it is to be questioned whether he continues to be the right kind of adviser.

MARKS. But you know that the subject of the present inquiry is whether Dr. Oppenheimer has been loyal, whether he can be trusted, whether he might not be a security risk?

TELLER. It wasn't I who proposed that these matters should be inquired into.

MARKS. Do you regard Dr. Oppenheimer as a security risk?

TELLER. His actions after the war appeared to me confused and complicated, and I personally would feel more secure if the vital interests of this country did not rest in his hands.

GARRISON. What do you understand by "security risk"?

TELLER. That there are well-founded doubts as to the discretion, the character, or the loyalty of a person.

GARRISON. In the sense of this definition, do you regard Dr. Oppenheimer as a security risk?

TELLER. No. . . . But I am no expert on questions of security.

MARKS. Do you think that his former left-wing sympathies affected his attitude in the matter of the hydrogen bomb?

TELLER. I think a man's attitude is always affected by his philosophy. But I do not know Dr. Oppenheimer well enough to analyze this.

MARKS. Could you tell us about Dr. Oppenheimer's philosophy?

TELLER. No. To me, it seemed contradictory. It surprised me

how effectively he had preserved his illusion that people, if patiently taught, might in the end achieve some political common sense. For instance, on the question of disarmament.

MARKS. You do not share that belief?

TELLER. I am convinced that people will learn political common sense only when they are really and truly scared. Only when the bombs are so big that they can destroy everything there is.

MARKS. If your advice should ever prove to have been wrong, would that disqualify you from serving the United States as a scientist?

TELLER. No, but I would no longer be the right man to occupy the leading position.

MARKS. Would you think it right if, because of it, your clearance were suspended?

TELLER. No.

MARKS. You know that Dr. Oppenheimer's clearance has been suspended until the final decision at these proceedings?

ROBB. I don't think it was suspended because he gave wrong advice!

MARKS. That is not what I said, Mr. Robb.

ROBB. But you are suggesting it, Mr. Marks.

MARKS. Dr. Teller, if it were up to you to grant Dr. Oppenheimer his clearance, would you do so?

TELLER. As I have no cognizance of possible reasons against it, yes, I would do so.

MARKS. These are my questions.

GRAY. Dr. Evans.

EVANS. What bothers me is this: is enthusiasm a good quality in a man who works on a weapons project?

TELLER. Without enthusiasm, we wouldn't have had the atom bomb in 1945, and neither would we have the hydrogen bomb.

EVANS. Good. Or, rather, not good, maybe. What I mean to

say is this: is enthusiasm just as good a quality in a man who has to advise the government?

TELLER. I don't know. You have already heard that I have a poor opinion of committees. I am not competent to judge. I only know that Dr. Oppenheimer would have helped us a great deal even if he had simply gone to Los Alamos and sat there in his office, twiddling his thumbs. He would have helped by the weight of his prestige alone.

EVANS. Should a man be reproached for not having shown any enthusiasm for a given thing—in our case, for the hydrogen bomb?

TELLER. No, he shouldn't. But one can state it as a fact, and one can inquire into the reasons for it.

EVANS. Have you never had any moral scruples about the hydrogen bomb?

TELLER. No.

EVANS. How did you manage to come to terms with that problem?

TELLER. I never regarded it as *my* problem.

EVANS. You feel that one can produce something, produce a hydrogen bomb, or something like that, and then say: "It isn't *my* problem what is to be done with it, run along and cope with it yourselves." Is that your attitude?

TELLER. It is not a matter of indifference to me, but I cannot possibly foresee the consequences, the full range of practical uses, which are part and parcel of an invention.

EVANS. Isn't it possible to foresee quite well the range of practical uses for a hydrogen bomb?

TELLER. No. It may be a good thing—and we all hope for it —if the hydrogen bomb were never used. And maybe in twenty or thirty years' time the face of the earth will be beneficially changed by the principle on which the bomb was developed—artificially generated solar energy, the cheapest and most powerful energy we have ever known.

EVANS. God grant that it be so, Dr. Teller.

TELLER. For instance, when Hahn in Germany first succeeded

in splitting uranium he never thought that the energy thus released might be used for producing explosions.

EVANS. Who was the first to think of it?

TELLER. Oppenheimer. And it was a constructive thought which only naïve people call immoral.

EVANS. Would you explain this to an older colleague?

TELLER. Discoveries in themselves are neither good nor evil, neither moral nor immoral, but merely factual. They can be used or misused. This applied to the internal-combustion engine, and it applies to nuclear energy. By painful experience, man has always learned in the end how best to use them.

EVANS. You say this although, judging from what you stated before, you have no great faith in common sense?

TELLER. I have faith in facts, which may in the end even produce common sense, occasionally.

EVANS. I have recently read in the newspapers about a terrible accident which was caused by one of our hydrogen bomb tests . . .

TELLER. Bikini?

EVANS. Yes, recently, twenty-three Japanese fishermen were seriously affected by radiation. One of them is already dead.

TELLER. Yes, that's right.

EVANS. How could such a thing happen?

TELLER. Their trawler drifted into a radioactive blizzard because the wind suddenly shifted from north to south, unfortunately.

EVANS. How did you react to this news about the fishermen?

TELLER. We set up a commission to study all the effects. And we were able to make considerable improvements in the meteorological forecasts for our tests.

EVANS. What kind of people are physicists?

TELLER. You mean: do they beat their wives, do they have hobbies, that kind of thing?

EVANS. I mean: do they differ from other people? I asked Dr. Oppenheimer the same question.

TELLER. What was his answer?

EVANS. That they are just like other people.

TELLER. Sure. They need a little bit more imagination and a little bit better brains, for their job. Apart from that, they are just like other people.

EVANS. I have been asking myself this question ever since I've been sitting on this Board. . . . Thank you very much.

GRAY. Perhaps you would like to ask Dr. Teller some questions, Dr. Oppenheimer?

OPPENHEIMER [*haughtily*]. No. [OPPENHEIMER *and* TELLER *look at each other for a moment.*] No.

GRAY. Then I would like to thank you for your evidence, Dr. Teller, which has touched upon some essential points, I think.

TELLER. May I make a general explanatory statement?

GRAY. Of course.

TELLER. I think it is necessary to say something about our problems. All great discoveries had, at first, a devastating effect on the state of the world and on its image in our minds. They shattered it and introduced new conditions. They forced the world to move forward. But this was possible only because the discoverers were not afraid of the consequences of their discoveries, no matter how terrifying these were to all those who wanted to preserve the world as it was, and hang a big notice on it, saying: Please do not disturb.

This was the case when the earth was discovered to be a planet among other planets; and this is the case now that we have been able to reduce matter, seemingly so complicated, to only a few elements which can be transformed and which release immeasurable energies.

If we persevere with our work, regardless of the consequences, we shall force man to adjust to these new energies, and to put an end to that state of the world in which he was half free, half enslaved! God alone knows whether this might not happen through a nuclear war—which would be terrible, like any other war, but, limited or unlimited, would not necessarily mean more suffering than in former wars, and which would probably be shorter though more violent.

If we shrink from the temporary aspect of discoveries, their powers of destruction—and I find this is the reaction of many a physicist—we'll get stuck half-way, and we'll be overwhelmed by the difficulties which our discoveries brought into the world.

Because of my relentless logic, many people regard me as an incorrigible warmonger, I know that, I read that in the newspapers; but I hope the time will come when I shall be called a peace monger, when the unspeakable horror of our destructive weapons will have discredited war as the classic means of realizing political aims, for ever.

EVANS. In case of survival, Dr. Teller, as they call it in the insurance business. But, supposing your prognosis turned out to be wrong, we perhaps ought to consider that mankind would have no chance of correcting the mistake. This is something new. Maybe even a physicist can't simply let it go at that.

TELLER. I don't think I do.

GRAY. Was that the explanatory statement you wished to make?

TELLER. Yes.

GRAY. Very well. Thank you.

TELLER *bows slightly to the Board and leaves the room.*

GRAY. The next witness is Dr. Bethe. . . . Has he arrived?

GARRISON. I'll have a look.

GARRISON *walks toward the door when the* OFFICIAL *enters with* BETHE. BETHE *stops in front of the witness stand.*

GRAY. Hans Bethe, do you swear that the testimony you are to give the Board shall be the truth, the whole truth, and nothing but the truth, so help you God?

BETHE. I do.

GRAY. Will you please take the stand. . . . When did you come to America, Dr. Bethe?

BETHE. In 1935. About the same time as Teller.

GRAY. Where did you originally come from?

BETHE. Munich. I spent a short time in England, and then I taught nuclear physics here, until I went to Los Alamos.

GRAY. You may question the witness, Mr. Garrison.

GARRISON. Were you in charge of the Theoretical Physics Department at Los Alamos?

BETHE. Until the end of the war, yes.

GARRISON. Did Dr. Teller work in your department?

BETHE. Yes.

GARRISON. How did you get on with him there?

BETHE [smiling]. Not at all. Edward Teller is a friend of mine, but it is very difficult to work with him.

GARRISON. Why?

BETHE. Edward is a highly gifted guy, full of brilliant ideas which he pursues with dogged fanaticism, until he drops them. Then he spends whole nights playing the piano, until he again gets some brilliant ideas, and he demands every time that one should be just as enthusiastic as he is himself. I am not saying this to belittle him, he is a genius, but he needs some other person to sort out his ideas for him. In the end it was better to forego his services than to have a whole team disrupted.

GARRISON. What do you mean by "forego his services"?

BETHE. We decided to relieve him of all the work on our own program because he was interested in nothing but the hydrogen bomb. Although we needed him badly.

GARRISON. Who took Teller's place?

BETHE. Klaus Fuchs.

GARRISON. Was there considerable tension between Dr. Teller and Dr. Oppenheimer at Los Alamos?

BETHE. They didn't like each other. Teller often complained that his work was not sufficiently appreciated, and that Oppenheimer was not sufficiently enthusiastic about it. But Oppenheimer had to co-ordinate a huge laboratory towards producing the atom bomb, and Edward was wrong to complain.

GARRISON. Was there any work done on thermonuclear development at Los Alamos?

BETHE. Oppenheimer had directed a whole group of people from my division to work on it—among them, Teller.

GARRISON. Did you work together well with Dr. Oppenheimer?

BETHE. I think we have been good friends ever since we met at the University of Göttingen in 1929.

GARRISON. Were there sufficient facilities for a hydrogen bomb program by the end of the war?

BETHE. Certainly not. Because I was coming here today, I checked with Fermi, to make sure. He was of the same opinion.

GARRISON. After the atom bomb test at Alamogordo, was there a plan to start a large-scale hydrogen bomb program?

BETHE. We discussed the possibilities. We realized we only had sufficient resources for an intensified research program. There was a plan for an intensified research program.

GARRISON. Dr. Teller has testified here that the plan for a hydrogen bomb program was abandoned only after Hiroshima, and that the reasons for abandoning it were the moral scruples of the scientists, in particular Dr. Oppenheimer. Is that right?

BETHE. No. The scientific ideas were frustrated because there were neither sufficient technical facilities nor sufficient personnel. . . . But it is true to say that Hiroshima made a great difference to us.

GARRISON. What effect did Hiroshima have on the physicists at Los Alamos?

BETHE. We had been working for several years under rigorous military conditions, and none of us had really stopped to think of those consequences. But Hiroshima put us face to face with the consequences—and, from then on, nobody could work on these weapons without being aware that they would actually be used.

GARRISON. Thereupon, what did you do?

BETHE. I left Los Alamos and went to Ithaca, where I taught physics. . . . I think it is well known that a number of scientists, including myself, appealed to the President and to the public, and I think it was the right thing to do.

GARRISON. Did you return to Los Alamos later on?

BETHE. Yes. After the outbreak of the war in Korea. I worked there until we tested the hydrogen bomb.

GARRISON. During that time, did you have moral scruples about working on the hydrogen bomb?

BETHE. Very strong ones. I still have them now. I have helped to create the hydrogen bomb, and I don't know whether it wasn't perhaps quite the wrong thing to do.

GARRISON. Why, then, did you go back to Los Alamos?

BETHE. The armaments race towards the H-bomb was in full swing, and I became convinced that we ought to be the first to have that dreadful weapon—if it could be produced at all. Yet, when I went to Los Alamos, I hoped that its production would prove impossible.

GARRISON. Dr. Teller has testified here that you wanted to take charge of the hydrogen bomb program at an earlier date, and it was due to Dr. Oppenheimer that you changed your mind. Is that right?

BETHE. I suppose Teller meant the talk we had with Oppenheimer, after the Russian atom bomb.

GARRISON. Before that visit, had you promised Teller that you would go to Los Alamos?

BETHE. I was undecided. On the one hand, I was very impressed by some of the ideas, and it was tempting to be able to work with the new calculating machines which had been released only for war projects. On the other hand, I was deeply troubled because I doubted that the H-bomb would solve any of our problems.

GARRISON. Did Dr. Oppenheimer speak against the hydrogen bomb?

BETHE. He presented us with facts, arguments, points of view.

He appeared to be just as undecided as I was myself. I was very disappointed.

GARRISON. Did you say to Teller afterwards that you would go to Los Alamos?

BETHE. Yes.

GARRISON. Why did you change your mind about it?

BETHE. Because I could not rid myself of my doubts. I spent a whole night talking with my friends, Weisskopf and Placzek, both of them eminent physicists, and we agreed that after a war with hydrogen bombs, even if we were to win it, the world would no longer be the world which we wanted to preserve, and that we would lose all the things we were fighting for, and that such a weapon should never be developed.

GARRISON. Is it known to you that Dr. Oppenheimer turned other physicists against the hydrogen bomb?

BETHE. No.

MORGAN. Why, then, did he make the General Advisory Committee's secret report accessible to the leading physicists who were working on the program, Dr. Bethe?

BETHE. Oppenheimer? He did it on instructions from Senator McMahon.

GARRISON. Would you tell us who is Senator McMahon?

BETHE. He headed the Senate Committee on Atomic Matters, and was one of the apostles of the H-bomb.

GARRISON. Do you think that because of Dr. Oppenheimer's attitude the development of the hydrogen bomb was decisively delayed, possibly for a matter of years?

BETHE. No. It was Teller's brilliant idea which made its production possible.

GARRISON. In Dr. Teller's opinion, you, or Fermi, or others, might have had the same brilliant idea if the program had been started earlier. Is that feasible?

BETHE. I don't know. I don't think the discovery of the Theory of Relativity, or something of that order, is made every day.

GARRISON. Why did Teller not manage to get a sufficient number of capable people for the program?

BETHE. Probably one of the reasons was the general uneasiness about it, and another reason was Teller himself. He is a wonderful physicist, but even his friends will have asked him: "All right, Edward, you'll get the acts together, but who is going to run the show?"

GARRISON. Did Dr. Oppenheimer oppose the crash program after the President had given the order to initiate it?

BETHE. After that, Oppenheimer only discussed how the H-bomb was to be made. Contrary to myself, who questioned its political usefulness.

GARRISON. What was his attitude to security requirements at Los Alamos?

BETHE. Many of us regarded him as too pro-government. That was my criticism too.

GARRISON. Dr. Bethe, you told us that you and Dr. Oppenheimer are good friends?

BETHE. Yes.

GARRISON. If Dr. Oppenheimer had to face a conflict of loyalties in which he would have to choose between you and the United States—to which loyalty, do you think, would he give the preference?

BETHE. Loyalty to the United States. I hope it will never come to that.

GARRISON [with a look at GRAY]. Thank you very much, Dr. Bethe.

GRAY looks questioningly at ROBB. ROLANDER makes a gesture to indicate that he wishes to cross-examine BETHE.

GRAY. Mr. Rolander.

ROLANDER. How long did Fuchs work in your division, Dr. Bethe?

BETHE. A year and a half.

ROLANDER. Did he work well?

BETHE. Very well.

ROLANDER. Did you ever notice that he behaved incorrectly in matters of security?

BETHE. No.

ROLANDER. Did you ever consider him a security risk?

BETHE. No.

ROLANDER. But it turned out that he had transmitted secret information to the Russians?

BETHE. Yes. . . . May I ask what you wish to imply?

ROLANDER. No, sir, since you are the witness, and not I! . . . When Dr. Teller visited you in Ithaca and asked you to take charge of the program, did you discuss the amount of your monthly salary?

BETHE. Yes. Teller made me an offer, and I asked for more.

ROLANDER. How much did you ask?

BETHE. Five thousand dollars a month.

ROLANDER. Did Teller agree to that?

BETHE. Yes.

ROLANDER. Does a man demand a higher salary when he is undecided whether to accept the job?

BETHE. In my case, yes. Good ideas are expensive. I like to eat well.

ROLANDER. I have an article here, from the periodical *Scientific American*; it dates back to the beginning of 1950. There you write: "Are we supposed to convince the Russians of the value of the individual by killing millions of them? If we fight a war with H-bombs, and win it, history will not remember the ideals we fought for but the method we used to enforce them. That method will be compared to the war techniques employed by Genghis Khan." Did you write that?

BETHE. I think it sounds sensible. The article was confiscated at the time because it was supposed to have given away important armaments secrets.

ROLANDER. You wrote that article a few weeks after you had turned down Teller's proposition?

BETHE. I think so.

ROLANDER. Yet, a couple of months later, you went to Los Alamos to develop the hydrogen bomb?

BETHE. Yes. . . . The article you've just read expresses the opinions I still have today.

ROLANDER. Your opinions today?

BETHE. Yes. We can justify the development of the hydrogen bomb only by preventing it being used.

ROLANDER. Thank you, Dr. Bethe.

GRAY. If I understand you correctly, you think it was wrong to develop the hydrogen bomb?

BETHE. I do.

GRAY. What should we have done instead?

BETHE. We ought to have come to an agreement by which nobody was allowed to produce this damned thing, and any breach of the agreement would mean war.

GRAY. Do you think there would have been the slightest chance of such an agreement in those days?

BETHE. Presumably it would have been easier to come by than to do the things we now have to do.

GRAY. What are you referring to?

BETHE. It seems the two power blocs haven't got much time left to decide whether to commit a double suicide with each other, or how to get that thing out of this world again.

GRAY [to ROBB]. Any further questions?

ROBB *shakes his head.* EVANS *raises his hand.*

EVANS. I would like to ask you something as an expert. Dr. Teller said here that a nuclear war, even if unlimited, would not necessarily mean more sufferings than in former wars. What is your opinion?

BETHE. That I can't bear to listen to such nonsense. I beg your pardon.

EVANS. That's all right.

GRAY. Thank you for having come here, Dr. Bethe.

BETHE. I considered it my duty. [*He rises to his feet.*] May I ask Dr. Oppenheimer to call me up at my hotel when he has finished here?

OPPENHEIMER. How long shall we still be here, Mr. Chairman?

GRAY. We have a big program behind us, we can recess until tomorrow . . . Mr. Robb?

ROBB. As Mr. Griggs is already waiting, I would like to ask for a few minutes for Mr. Griggs.

MARKS. In that case, may we also hear Dr. Rabi?

GRAY. Very well . . . Would you call Dr. Griggs, please?

An OFFICIAL *leaves the room to fetch* GRIGGS.

OPPENHEIMER [*to* BETHE]. We might eat together.

BETHE. Fine.

BETHE *leaves the room. Immediately afterward,* GRIGGS *enters.*

GRAY. Dr. Griggs, do you wish to testify under oath?

GRIGGS. Yes. My name is simply Griggs, David Tressel Griggs.

GRAY. David Tressel Griggs, do you swear that the testimony you are to give the Board shall be the truth, the whole truth, and nothing but the truth, so help you God?

GRIGGS. I do.

GRAY. Would you please take the stand. . . . Was it your own wish to testify here as a witness?

GRIGGS. I was ordered by the Air Force to come here.

GRAY. I should point out to you that you are only allowed to give your own personal views here.

GRIGGS. Of course.

GRAY. What is your present position?

GRIGGS. Chief Scientist of the Air Force.

GRAY. What is your special field?

GRIGGS. Geophysics.

GRAY. The witness may now be questioned.

ROBB. I would like to ask you, Mr. Griggs, whether you know Dr. Oppenheimer's attitude in the matter of the hydrogen bomb?

GRIGGS. Yes. We saw all the estimates and reports submitted by him.

ROBB. How would you describe his attitude?

GRIGGS. In the course of observation and analysis, I finally came to the conclusion that there was a silent conspiracy among some prominent scientists, a conspiracy directed against the hydrogen bomb. This group endeavored to prevent or delay the development of the hydrogen bomb, and it was led by Dr. Oppenheimer.

ROBB. Is this your private opinion, or is it shared by others?

GRIGGS. It is my own opinion, and it is the opinion also of Mr. Finletter, the Secretary of the Air Force, and of General Vandenberg, the Chief of Staff of the Air Force.

ROBB. What were the facts which convinced you?

GRIGGS. For a long time I was puzzled by the actions of Dr. Oppenheimer and some others. Then, one day, I found a clue to it all.

ROBB. When was that?

GRIGGS. In 1951, there was a conference on strategy, on the so-called Vista Project. The subject of the discussions was whether the emphasis should in future be on a strategic H-bomber fleet, or on a strengthening of the air defenses, the warning systems, anti-aircraft missiles, and so forth. A kind of electronic Maginot Line, purely defensive and very costly. The Air Force had decided on an H-bomber fleet.

ROBB. And Oppenheimer?

GRIGGS. I am coming to that. . . . One day, when the two opposing parties were not yet clearly defined, and when I had just attacked those who were in favor of air defense, Dr. Rabi went to the blackboard and wrote the word "ZORC" on it.

ROBB. ZORC? What is the meaning of "ZORC"? Would you spell it, please.

GRIGGS. Z-O-R-C. These are the initials of a group consisting of Zacharias, Oppenheimer, Rabi, and Charlie Lauritzen. They advocated world disarmament.

ROBB. Why did Rabi write this word on the blackboard?

GRIGGS. In my opinion, in order to indicate to his supporters which side they were to take in the discussions at the conference.

MARKS. Mr. Robb, may I ask the witness a question?

ROBB. You will have an opportunity to ask Mr. Griggs any question you like when you cross-examine him, Mr. Marks! [*To* GRIGGS.] How did the conference end?

GRIGGS. On three major points, the recommendations of the conference were entirely opposed to the line taken by the Air Force, and this part of the recommendations had been written by Dr. Oppenheimer himself.

ROBB. Have you observed other activities of the Oppenheimer group?

GRIGGS. Yes. A story was being spread among the scientists that Mr. Finletter had said in the Pentagon: "If we had such-and-such a number of hydrogen bombs, we could rule the world." . . . This was meant to prove that we had irresponsible warmongers at the head of the Air Force.

ROBB. Did you speak to Dr. Oppenheimer about it?

GRIGGS. Yes. I challenged him, and asked whether he himself had told that story. He said he had heard the story but didn't take it seriously. I said that I myself took it very seriously indeed because it was designed to spread lies, for a definite purpose. Dr. Oppenheimer asked me whether this meant that I doubted his loyalty, and I said, Yes, that was exactly what I meant.

ROBB. What was his reaction to this?

GRIGGS. He said that I was a paranoiac, and simply walked away! Later I understood well enough why Dr. Oppenheimer, in Princeton, praised the technical side of the program but boycotted the second laboratory although the Air Force was willing to provide the money for it. I understood the obstacles which Teller complained about. Particularly when I had read the F.B.I. report.

ROBB. Do you think there is a connection between his left-wing associations and his attitude in the matter of the hydrogen bomb?

GRIGGS. I am convinced of that.

ROBB. Do you regard Dr. Oppenheimer as a security risk?

GRIGGS. Yes, a very serious one.

ROBB. Thank you, Mr. Griggs.

GRAY. Mr. Marks?

OPPENHEIMER *makes an emphatically negative gesture to* MARKS.

MARKS. Dr. Oppenheimer wishes his counsel to refrain from cross-examining Mr. Griggs. The members of the Board will not construe this as an indication of agreement.

GRAY. Any questions to Mr. Griggs? . . . Yes, Dr. Evans.

EVANS. Were there many people present when Rabi wrote those four letters, the secret code, on the blackboard?

GRIGGS. Quite a number.

EVANS. And they all saw it?

GRIGGS. Yes. They reacted to it.

EVANS. In what way?

GRIGGS. In various ways. Some of them laughed.

EVANS. Mr. Griggs, if you belonged to a group of conspirators, and you wanted to communicate something to your fellow conspirators, would you think it wise to write it on a blackboard?

GRIGGS. I did not say it was wise. I have never taken part in a conspiracy myself.

EVANS. Neither have I, but I think I'd rather have gone to my confederates and said to them: we'll do this in such-and-such a way.

GRIGGS. The one doesn't exclude the other. The fact is: Rabi wrote ZORC on the blackboard.

EVANS. You said so, yes.

GRAY. Any further questions? . . . May I thank you for having come here, Mr. Griggs. [GRIGGS *bows stiffly to* GRAY, *and leaves the room.*] We shall now call Dr. Rabi.

MARKS. I believe he is already here.

> DR. I. I. RABI *enters and takes the witness stand.*

GRAY. Do you wish to testify under oath, Dr. Rabi?

RABI. Certainly.

GRAY. I must ask your full name.

RABI. Isador Isaac Rabi.

GRAY. Isador Isaac Rabi, do you swear that the testimony you are to give the Board shall be the truth, the whole truth, and nothing but the truth, so help you God?

RABI. I do . . . [*Wryly.*] . . . and I know that the penalties for perjury are dire.

GRAY. You may question the witness, Mr. Marks.

MARKS. Dr. Rabi, what is your present occupation?

RABI. I am the Higgins professor of physics at Columbia University.

MARKS. What official positions do you have with the government?

RABI. Let me see if I can add them all up. Chairman of the General Advisory Committee, member of the Scientific Advisory Committee which is supposed in some way to advise the President, member of a whole lot of boards, research development laboratories, and so on. I added them up once and it amounted to 120 working days last year, so you might well ask what time I spend at Columbia.

MARKS. I won't take up much of your time.

RABI. If I get away from here early enough, I must see Chairman Strauss of the Atomic Energy Commission.

MARKS. On behalf of Dr. Oppenheimer?

RABI. Yes, certainly. But even more on behalf of U.S. security. To tell you frankly, I have very grave misgivings as to the nature of this charge and the general public discussion it has aroused. Important security information absolutely vital to the United States may bit by bit inadvertently leak out. I myself am confused about what is classified and what is not.

GRAY. I should point out, Dr. Rabi, that we regard the proceedings as a matter confidential in nature between the Commission and Dr. Oppenheimer. The Commission will make no public release of matters pertaining to these proceedings.

RABI. I hope not. It makes your hair stand on end to hear high officers and people in Congress say some of the things they say. A man like Oppenheimer knows when to keep his mouth shut. But the papers and magazines skirt around very important security information. Bits and pieces can be put together. I am very much worried about that.

GRAY. There will be no public release of matters pertaining to these proceedings.

A cover of the hearing transcript and a headline bearing date of publication is projected on the screens.

MARKS. Now, Dr. Rabi, how long have you known Dr. Oppenheimer?

RABI. Since 1928 or '29. We worked together all through the war.

MARKS. Do you know him intimately?

RABI. Yes, whatever the term may mean. I think I know him quite well.

MARKS. Would you describe for us what took place in 1949 with the General Advisory Committee regarding going ahead with the thermonuclear, the H-bomb program?

RABI. As I recollect it now—it is five years ago—Dr. Oppenheimer, who was then Chairman, called us together and started very solemnly to consider, *not* whether we should make the bomb, but whether there should be a crash program to develop what was then a very vague thing. Different people had different thoughts on it: I myself took the dimmest technical view. Others were more optimistic, but technically we really did not know what we were talking about. . . . Also, there was the military and political question.

MARKS. So the General Advisory Committee did not deal solely with technical aspects of the program?

RABI. There were technical, military, and the combination of military-political questions. We felt—I am talking chiefly about myself—that to protect American lives was worth anybody's while, but that a crash program for an H-bomb should not be at the expense of continental defense. First we have the H-bomb, then they have it; you can't just go in and slug it out with a punching arm like the H-bomb with no defense guard. And politically, this was not just a weapon, it was very much more. We felt it was essential, but a crash program was just a sort of horseback thing. We didn't even know whether this thing contradicted the laws of physics.

MARKS. You didn't know what?

RABI. Whether it contradicted the laws of physics.

MARKS. In other words, it could have been altogether impossible?

RABI. It could have been altogether impossible . . . the thing we were talking about. I want to be specific.

MARKS. You have spoken of differences of opinion. Could you explain?

RABI. One group felt that this projected weapon was just no good as a weapon, not technically, but militarily. The possible targets were very few in number and so on. Dr. Fermi and I felt that the whole discussion raised an opportunity for President Truman to make some political gesture which would strengthen our moral position should we decide to go ahead with it.

GRAY. With respect to the development of the hydrogen bomb and the issue of who was for and who was against, was it your impression that Dr. Oppenheimer was unalterably opposed to the development?

RABI. No. I distinctly remember Dr. Oppenheimer saying he would be willing to sign statements by both groups.

GRAY. Both groups?

RABI. Yes. There was no difference as far as a crash program was concerned. Both groups thought that was not in order.

GRAY. Subsequent to the President's decision to go ahead with the bomb, would you say Dr. Oppenheimer encouraged the program and assisted with it?

RABI. Yes, sir.

EVANS. Did you think it was appropriate for the General Advisory Committee to speak about these rather nontechnical but more political, diplomatic, and military considerations?

RABI. It would be very hard for me to tell you now why we thought it was appropriate, but we thought so.

MARKS. If you can, Dr. Rabi, what was the connection between the reluctance you have just described at Los Alamos in 1949 and a later meeting in 1951?

RABI. That was an entirely different meeting. At that meeting we really got on the beam because a new invention had occurred.

MARKS. Are you referring to Dr. Teller's discovery?

RABI. I wouldn't call it Dr. Teller's discovery. Dr. Teller had a very important part in it, but I would not make a personal attribution.

MARKS. Do you believe Dr. Oppenheimer was responsible for any delay between 1949 and 1951?

RABI. There has been all this newspaper stuff about delay. Dr. Oppenheimer and I first discussed this project as early as 1943, but it wasn't until 1951 we had a situation you could really talk about, know what to calculate, and so on.

MARKS. Are you familiar with the term "ZORC," Dr. Rabi?

RABI. Isn't everyone? *Fortune* magazine gave that term some notoriety.

MARKS. When was that in relation to the Vista Conference?

RABI. I don't remember exactly. Several months before, I know that, because Zacharias made a joke of it at that meeting by writing it on the blackboard.

MARKS. Zacharias? Who was he?

RABI. Scientific Adviser to the Navy, a first-class nuclear physicist.

MARKS. It has been testified here that it was you, Dr. Rabi, who wrote "ZORC" on the board.

RABI. Not at all. It was Zacharias. Everyone had a good laugh at the expense of *Fortune*.

MARKS. Are you sure the article mentioning ZORC was published *before* the conference?

RABI. Weeks before. And they didn't mention it. They *invented* it.

MARKS. Dr. Rabi, this Board has the function of advising the Commission on Dr. Oppenheimer's security clearance. Do you feel you know Dr. Oppenheimer well enough to comment on this issue?

RABI. Dr. Oppenheimer is a man of upstanding character. He is a loyal individual, not only to the United States, which of course goes without saying in my mind, but also to his friends and to his organizations. He is a very upright character, a very upright character.

MARKS *indicates he is finished.*

ROBB. Dr. Rabi, you said you and Dr. Oppenheimer had your first discussion about the H-bomb in 1943?

RABI. From our very first contact at Los Alamos.

ROBB. And from 1943 on—even after the President's directive in 1950 that the bomb should be built—the progress was . . . negligible?

RABI. It was just a ball of wax. The President says go do something that nobody knows how to do. Just because he says something should appear at this end, doesn't mean it's going to appear just like that.

ROBB. Is it not true that while you and Dr. Oppenheimer were opposed to the crash program, other scientists were enthusiastic about it?

RABI. Oh yes. They were all keyed up to go bang into it.

ROBB. But you and Dr. Oppenheimer resisted.

RABI. Not the H-bomb as it exists. The thing that Oppenheimer and I—and others—were opposed to crash is not the

H-bomb as it now is; it was something that has not been made, probably never will be made, and we don't know to this day whether the thing would function anyway.

ROBB. Why did the development of something functional take so long?

RABI. The human mind! It takes time to get rid of ideas that were, and probably are, no good.

ROBB. Dr. Rabi, you have read the F.B.I. file on Dr. Oppenheimer?

RABI. I may say that the record is not something I wanted to see.

ROBB. No, I understand that.

RABI. In fact, I disliked the idea extremely of delving into the private affairs in this way of a friend of mine.

ROBB. Certainly. Doctor, did it surprise you to learn that the man you have here, today, called "upright" and "loyal" had lied to the security authorities when there was serious suspicion of espionage?

RABI. It did surprise me at the time. I thought he behaved foolishly.

ROBB. If you had been in Dr. Oppenheimer's place, would you have lied to the security authorities—the same as he did?

RABI. The Lord alone knows.

ROBB. But I am asking *you*.

RABI. I don't think I would have done anything more than Oppenheimer did unless I thought the man who contacted me was just a poor jackass and didn't know what he was doing.

GRAY. I should like to ask another question, Dr. Rabi. As of today, would you expect Dr. Oppenheimer's loyalty to the country to take precedence over loyalty to an individual or some other institution?

RABI. I just don't think that anything is higher in his mind or heart than loyalty to his country. I must say I think that our generation, Dr. Oppenheimer's and my other friends, created American physics . . . that between the years

1929 and, say, 1939 we made it to the top of the heap. And it wasn't just because certain refugees came out of Germany. But because of what we did here. This was a conscious motivation. And Oppenheimer's school of theoretical physics was a tremendous contribution. I don't know how we could have carried out the scientific part of the war without the contributions of the people who worked with Oppenheimer at Los Alamos. It was really a miracle of a laboratory, just a miracle of a place.

GRAY. Would you expect Dr. Oppenheimer today to follow the course of action he followed in 1943 with regard to the Chevalier incident?

RABI. At the present time if a man came to him with a proposal like that, I think Oppenheimer would clamp him into jail.

GRAY. You are saying that in your judgment Dr. Oppenheimer has changed?

RABI. He has learned. He was always a loyal American. There was no doubt in my mind as to that. But he has learned more about the way you have to live in the world as it is now.

EVANS. Let me ask you a question that has nothing particularly pertinent to the proceedings. We have a mutual friend, George Pegram. I wonder if you know if he is still active?

RABI. Wonderfully. He is seventy-eight, doing two men's work.

EVANS. I wish you would tell him Dr. Evans asked about him.

RABI. I would be delighted to.

EVANS. Thank you.

MARKS. Dr. Rabi, would you be confident or would you not be confident that today Oppenheimer would resolve the question of his responsibility to the country or the public in a way you would?

RABI. I think he would be very conscious of his position not to impair his usefulness to the United States. I think he is just a much more mature person than he was then.

MARKS. That is all.

GRAY. You understand, Dr. Rabi, that these proceedings are not in the nature of a trial? And that we do not bring a verdict.

RABI. But your opinion will have the same weight. I think this business of suspending Dr. Oppenheimer's clearance is very unfortunate. It doesn't seem to me the kind of thing called for against a man who accomplished what he accomplished. We have a whole series of A-bombs. This is just a tremendous achievement. What more do you want? Mermaids? If the end of the road is this kind of hearing—which can't help but be humiliating—then where have we come?

GRAY. Let me thank you for your views, Dr. Rabi. And for taking the time to come here today.

RABI exits.

Change of lighting. The hangings close.

SCENE 3

The following text is projected on the hangings:

IN THE MORNING OF MAY 6, 1954, THE COMMISSION CONCLUDED THE INTERROGATION OF WITNESSES. FORTY WITNESSES HAD TESTIFIED IN THE MATTER OF J. ROBERT OPPENHEIMER. THE RECORD OF THE HEARINGS RAN TO 3,000 TYPEWRITTEN PAGES.

THE INTERROGATION OF THE WITNESSES WAS FOLLOWED BY THE SPEECHES FOR THE PROSECUTION AND FOR THE DEFENSE.

GRAY. I call on Mr. Robb to place his summing-up before the Board. The same right is granted to counsel for the defense, and Mr. Marks will avail himself of that right. Thereafter, the Board will recess in order to reach a final decision, as it has been instructed to do. Mr. Robb.

ROBB. Mr. Chairman, members of the Board. In the course of the three and a half weeks during which Dr. Oppenheimer has been before us, there has been recorded here the life story of an eminent physicist, the contradictions, the conflicts, and I confess I have been moved by it. I feel its tragic aspect. None of us has any doubt of Dr. Oppenheimer's great merits, and only a few would be able to resist the charm of his personality. However, it is our arduous duty to examine whether the safety of this country, in a field as important as nuclear energy, rests secure in his hands. Unfortunately, our security is at present threatened by the Communists who aim at establishing their form of government all over the world. According to his own testimony, Dr. Oppenheimer has for a long period of his life been in such close sympathy with the Communist movement that it is difficult to say in what respect he actually differed from a Communist. His nearest relatives, the majority of his friends and acquaintances, were either Communists or fellow travelers. He attended Communist meetings, he read Communist newspapers, he donated sums of money and belonged to a great number of camouflaged Communist organizations. I have no doubt that, to begin with, he was impelled by admirable motives, by a desire for social justice, by a dream of an ideal world. However, in the course of these proceedings, my conviction grew that Dr. Oppenheimer never abandoned his Communist sympathies, even when his enthusiasm had cooled, even when, disappointed, he turned away from the political manifestations of Communism in Russia.

His Communist sympathies were apparent when Communist physicists, recommended by Dr. Oppenheimer, gained key positions in work on secret war projects; they were apparent when, using his very considerable influence, he had these men retained on the war projects even though they were distrusted; and they finally were apparent in the Eltenton-Chevalier incident when he hesitated for half a year before he reported a serious suspicion of espionage, and when he deliberately lied to the security authorities, and when he placed his loyalty to a Communist friend above his loyalty to the United States.

It has been argued here that this incident belongs to the past, and that Dr. Oppenheimer has proved his absolute loyalty by his great services in connection with the atom bomb. I cannot share this view, although I do not dispute his merits in relation to Los Alamos. Rather, I see in his actions after the war, and particularly in the matter of the hydrogen bomb, the same manifestation of the same old Communist sympathies.

According to many testimonies, Dr. Oppenheimer was equally enthusiastic about the atom bomb *and* the hydrogen bomb as long as the enemy was Nazi Germany. But when it became clear that there were not only right-wing but also left-wing dictatorships that threatened us, and when Russia became our potential enemy, his scruples about the hydrogen bomb grew ever stronger, and he advocated internationalization of nuclear energy although it had been due entirely to our atom-bomb monopoly that we succeeded in stopping the Russians in Europe and Asia.

According to his own testimony, Dr. Oppenheimer felt profoundly depressed when it proved impossible to come to any agreement with the Russians. But he did not draw the conclusion from it which would have been in the best interests of the United States: the conclusion that the development of the hydrogen bomb should be stepped up before the Russians had their own atomic bomb.

Even when our danger was brought home to us in no uncertain manner by the existence of the Russian bomb, he used his considerable influence to oppose the crash program for the hydrogen bomb, and he yet again advocated negotiations with Soviet Russia in order to prevent the development of such a weapon. When, in spite of it, the order was given to initiate the crash program, and although he was fascinated by the brilliant new scientific ideas for the development of the hydrogen bomb, he still persevered in recommending a long-term research program; and when the date had already been fixed for the first test, he tried to postpone it in order not to jeopardize any disarmament negotiations.

We have heard several witnesses stating here that they found the discrepancy between his words and his actions inexplicable. Some of them—for instance, Major Radzi, William Borden, and Mr. Griggs—had come to the conclusion that a particularly subtle kind of treason lay at the root of it. But whoever had the opportunity to observe Dr. Oppenheimer for three and a half weeks, as we did, and whoever is as impressed with his personality as we are, will realize that this man is no common traitor. I am convinced that Dr. Oppenheimer wanted to serve the best interests of the United States. But his actions after the war, his obvious failure in the matter of the hydrogen bomb, were in fact injurious to the interests of this country—indeed, according to Dr. Teller's most convincing exposition, we could already have had the hydrogen bomb four or five years earlier if Dr. Oppenheimer had supported its development.

What is the explanation for such a failure in a man so wonderfully gifted, a man whose diplomatic acumen, whose sagacity have been praised here so often?

This is the explanation: Dr. Oppenheimer has never entirely abandoned the utopian ideals of an international classless society. He has kept faith with them consciously or unconsciously, and this subconscious loyalty could only in this way be reconciled with his loyalty to the United States. It is in this contradiction that his tragedy lies, and it is a lasting tragedy which prevents him from serving the best interests of the United States in that difficult sphere—even though he honestly wishes to do so. This is a form of treason which is not known in our code of law; it is ideological treason which has its origins in the deepest strata of the personality and renders a man's actions dishonest, against his own will.

I speak of lasting tragedy because, in the course of these proceedings, Dr. Oppenheimer has never once availed himself of the opportunity to dissociate himself from his former political ideas and from his Communist associations. On the contrary, he has kept up these associations throughout

the war, and after, and still has some of these personal associations even now. Neither has he ever recognized how
wrong his actions have been, and he has not regretted them.
And when we heard him saying here that, in the age of
nuclear energy and weapons of mass destruction, the world
needs new forms of human, economic, political co-existence
—then I see in those words also a projection of his old ideals.

What America in fact needs today is a strengthening of
her economic, military, and political power.

We have reached a point in our history when we must
recognize that our freedom has its price, and historical
necessity does not permit us to give any discount to anybody, even if he is a man who has rendered great service.
This does not mean that we forget his past services; indeed,
we respect them. It is my conviction, taking all facts into
account, that Dr. Oppenheimer is *not* eligible for security
clearance.

GRAY. Thank you, Mr. Robb. The Board will now hear counsel for the defense. Mr. Marks.

MARKS. Mr. Chairman, Dr. Evans, Mr. Morgan. Mr. Robb
has spoken of the great merits and the tragic aspects of my
client. I take this sympathy to be an admission that no facts
have emerged in these proceedings which would impugn
Dr. Oppenheimer's loyalty.

It is generally known that in the thirties Dr. Oppenheimer had strong leanings towards radical left-wing and
Communist ideas, that he had friends who were Communists, and that he belonged to some organizations which
had Communist sympathies. In those days, this was the attitude of many, if not most, intellectuals; and their social
criticism corresponded to our policy of the New Deal,
which introduced a greater measure of social justice in our
country. What we have learned here about Dr. Oppenheimer's associations was already to be found in the ques-
tionnaires which Dr. Oppenheimer filled in before he
started working on secret war projects, and those facts were
fully known to the high and highest committees which
granted Dr. Oppenheimer his clearance in 1943 and 1947.

The material which the F.B.I. collected about Dr. Oppenheimer, and to which we have been denied access, was also known to the relevant committee already in 1947. I presume Mr. Robb would not have hesitated to acquaint us with that material if it had contained well-founded charges which are unknown to us.

Likewise, the security authorities had full cognizance of Dr. Oppenheimer's conduct in the matter of Eltenton-Chevalier. That matter was cleared up before, and Mr. Robb has failed to present us with any fresh evidence. There was no conflict of loyalties, since Dr. Oppenheimer regarded Chevalier as innocent and Chevalier in fact proved to be innocent. Although it was realized that there had been no such thing as attempted espionage, Dr. Oppenheimer did not hesitate to say that his conduct had been foolish, and nobody will doubt that his conduct today would be different from what it was in 1942.

There remains the question whether Dr. Oppenheimer endangered the security of the United States by opposing the hydrogen bomb program, against his better judgment, and with disloyal intent. It is not a matter of whether his advice was good or bad, but whether it was honest advice, and whether it was given in the best interests of the United States, or not.

Many experts have here testified to their opinion that Dr. Oppenheimer's advice was good advice when he submitted that the development of the hydrogen bomb should be prevented by means of an international agreement. He anticipated, and apprehended, the balance of fear which paralyzes us today. But other experts, such as Teller and Alvarez, had different views, and they carried their point. They sharply criticized his proposals, but even the determined advocates of the hydrogen bomb did not doubt that he had wanted to serve the best interests of America with his advice. Dr. Teller has complained here that Dr. Oppenheimer had not been sufficiently enthusiastic about the hydrogen bomb and that it was his lack of enthusiasm which delayed the hydrogen bomb for a matter of years. But how is a man to feel enthusiastic when he fears that such a weapon may in fact weaken America and imperil our whole civilization? How is a man to feel enthusiastic when he is confronted with a program which is technically not feasible and which is a program for a weapon that has all the strategic and political arguments against it? What would Dr. Teller say if he were reproached for his lack of enthusiasm for the atomic bomb during the war, and if it were pointed out to him that his place had therefore been taken by Klaus Fuchs, and thus it was his, Teller's, fault that atomic secrets had been given away? He would be quite right to consider this absurd, and the myth about the delay of the hydrogen bomb because of Dr. Oppenheimer's lack of enthusiasm is equally absurd.

It was in the light of his best judgment that Dr. Oppenheimer expressed an adverse opinion on a bad crash program. He was in complete agreement there with the foremost experts in the country. When, in spite of it, the order was given to initiate the crash program and when new ideas for the hydrogen bomb made its production feasible, he no longer discussed its political usefulness but supported the program to the best of his abilities. I do not see how any man's attitude could be *more* correct and *more* loyal than that.

Where are those dishonest actions which contradict his words? Where are the facts to bear out the allegation that Dr. Oppenheimer has been disloyal, that he cannot be trusted, that he endangers the security of the United States? Is the "closed Communist meeting" alleged by Mr. Crouch, is the "silent conspiracy" alleged by Mr. Griggs, such a fact? Was it traitorous conduct when Dr. Oppenheimer did not toe the line of Air Force enthusiasts in the weapons rivalry over the defense of the country? Dr. Oppenheimer had to advise the American government, and not the Air Force. He had to think of America, and not of the priority of one kind of weapon over another.

We may doubt the wisdom of his advice, and this is quite in order if his advice is no longer required, but we cannot cast doubt upon the loyalty of a man because we doubt the wisdom of his advice.

If we were to follow Mr. Robb's suggestion and introduced here the concept of ideological treason, a category which is not known in our code of law, we would destroy not only the scientific career of a great American but would destroy also the very foundations of our democracy.

I agree with Mr. Robb when he says that freedom has its price. Dr. Oppenheimer himself, in a newspaper article he wrote in defense of one of his colleagues, had the following to say about the price of freedom:

"Political opinion, no matter how radical or how freely expressed, does not disqualify a scientist from a high career in science; it does not impugn his integrity nor his honor. We have seen in other countries criteria of political orthodoxy applied to ruin scientists, and to put an end to their work. This has brought with it the attrition of science. It has been part of the destruction of freedom of inquiry, and of political freedom itself. This is no path to follow for a people determined to stay free."

GRAY. Thank you, Mr. Marks. The Board will now recess. We shall let you know the date of the final session. I thank all those present for their participation. In particular, I thank Dr. Oppenheimer.

OPPENHEIMER. Thank you, sir. [*The lighting changes.* OPPENHEIMER *steps forward to the footlights. The hangings close.*] On May 14, 1954, a few minutes to ten, physicist J. Robert Oppenheimer entered, for the last time, Room 2022 of the Atomic Energy Commission in Washington in order to hear the final decision of the Board, and to make a concluding statement. [OPPENHEIMER *returns to the stage itself.*]

SCENE 4

The following text is projected on the hangings:

THE FINAL DECISION

The members of the Board, OPPENHEIMER, *and counsel for both parties occupy their usual places.* GRAY *takes a report from a folder and rises to his feet to read it aloud.*

GRAY. After due consideration of all the facts the majority of this Board—the members Thomas A. Morgan and Gordon Gray jointly diverging from the conclusions reached by the member Ward V. Evans—propose to submit to the Atomic Energy Commission their final decision in the matter of J. Robert Oppenheimer, as follows:

"Although we consider Dr. Oppenheimer's numerous Communist associations in the past a grave indictment, and although Dr. Oppenheimer made the regrettable decision to continue some of those associations up to the present day, we find no indication of disloyalty as far as his present associations are concerned.

"Dr. Oppenheimer's attitude in the Eltenton-Chevalier incident appears to us a weightier matter. There, in a serious case of suspected espionage, he deliberately lied to the security authorities in order to protect a friend whose Communist background was well known to him, and thus he placed himself outside the rules which govern the conduct of others. It is not important whether it was in fact a case of attempted espionage; what is important is that he believed in the possibility of it being such. The continued falsification and false statements point to disquieting defects of character.

"Loyalty to our friends is one of the noblest qualities. But if a man puts his loyalty to friends above what may reasonably be considered his duties to his country and its security system, this is, beyond a doubt, incompatible with the interests of his country.

"As to Dr. Oppenheimer's attitude in the matter of the hydrogen bomb, we find it ambiguous and disquieting. If Dr. Oppenheimer had given the program his enthusiastic support, organized endeavor would have begun sooner, and we would have had the hydrogen bomb at a considerably earlier date. This would have increased the security of the United States. We believe that Dr. Oppenheimer's negative attitude in the matter of the hydrogen bomb stemmed from his strong moral scruples, and that his negative attitude had an adverse effect on other scientists. Although we

have no doubt that he gave his advice with loyal intent and
to the best of his ability, we note a deplorable lack of faith
in the United States government, a lack of faith which is
exemplified by his endeavor to prevent the development of
the hydrogen bomb by means of international agreements,
and is further exemplified by his demand for a guarantee
that we shall not be the first to use that weapon.

"We find that his conduct gives rise to considerable
doubt as to whether his future participation in a national
defense program would be unequivocally compatible with
the requirements of the country's security, if he persisted in
such conduct.

"Summing up our doubts, and with reference to evidence
of his basic defects of character, we conclude that Dr. Oppenheimer can no longer claim the unreserved confidence
of the government and of the Atomic Energy Commission,
a confidence which would have found expression in the
granting of his security clearance.—Gordon Gray. Thomas
A. Morgan."

Appendix by Gordon Gray:

"In my view, it would have been possible for us to reach
a different conclusion if we had been permitted to judge
Dr. Oppenheimer independently of the rigid rulings and
criteria now enforced upon us."

I now call on Dr. Evans to read his minority report.

GRAY *sits down.* EVANS *picks up a sheet of paper and holds it
close to his eyes in order to read it aloud.*

EVANS. "After due consideration of the facts laid here before
us, I regard Dr. Oppenheimer as absolutely loyal, I do not
consider him a security risk, and I see no reason why his
security clearance should be withheld.

"My arguments are as follows: Dr. Oppenheimer's former
Communist associations, including also his attitude to Chevalier, go back to a time *before* the great services he
rendered the United States. Dr. Oppenheimer has never
made a secret of those associations, and all the charges
brought here against him were already known when, lastly

in 1947, he was granted his security clearance. I am troubled by the phenomenon that an assessment of the *same* facts should change when there is a change in the political climate.

"In the discussions about the hydrogen bomb, it was not only Dr. Oppenheimer's right but his duty to express his own opinion. His views in this complicated matter were well founded; they coincided with the views of many of the foremost experts, and it is by no means certain whether his advice was not the best advice after all. When we inquire into a man's loyalty, however, it is not a matter of whether his advice was sound but whether it was honest. Moral and ethical reservations about the development of a weapon are not necessarily injurious to the interests of America, and it is common sense to consider in good time the consequences of a development which is fraught with so many possible repercussions.—Ward V. Evans."

GRAY. It thus is evident that the majority of the Personnel Security Board of the Atomic Energy Commission recommend that Dr. Oppenheimer shall not be granted his security clearance. [*To* OPPENHEIMER's *counsels.*] You may request the Atomic Energy Commission to review that recommendation. I now give Dr. Oppenheimer the opportunity he requested for a few concluding words. Dr. Oppenheimer.

OPPENHEIMER *rises to his feet.*

OPPENHEIMER. When, more than a month ago, I sat for the first time on this old sofa I felt I wanted to defend myself, for I was not aware of any guilt, and I regarded myself as the victim of a regrettable political conjunction. But when I was being forced into that disagreeable recapitulation of my life, my motives, my inner conflicts, and even the absence of certain conflicts—my attitude began to change. I tried to be absolutely frank, and this is a technique one must relearn when one has not been frank with people for many years of one's life. As I was thinking about myself, a physicist of our times, I began to ask myself whether there had not in fact been something like ideological treason, a category of treason Mr. Robb proposed should be considered here. It has become a matter of course to us that even basic research in the field of nuclear physics is top secret nowadays, and that our laboratories are financed by the military and are being guarded like war projects. When I think what might have become of the ideas of Copernicus or Newton under present-day conditions, I begin to wonder whether we were not perhaps traitors to the spirit of science when we handed over the results of our research to the military, without considering the consequences. Now we find ourselves living in a world in which people regard the discoveries of scientists with dread and horror, and go in mortal fear of new discoveries. And meanwhile there seems to be little hope that people will soon learn how to live together on this ever smaller planet, and there is little hope that, in the near future, the material side of their lives will be enhanced by the new beneficent discoveries. It seems a thoroughly utopian idea that atomic energy, which can be produced everywhere equally easily and equally cheaply, would be followed by other benefits for all; and that the electronic brains, originally developed for the great weapons of destruction, would in future run our factories and thus restore the creative quality to man's work. Our lives would be enriched by material freedom which is one of the prerequisites of happiness, but such hopes are not borne out by the reality we have now to live with. Yet they are the alternatives to the destruction of this earth, a destruction we fear and are unable to imagine. At these crossroads for mankind we, the physicists, find that we have never before been of such consequence, and that we have never before been so completely helpless. As I was looking at my life here I realized that the actions the Board hold against me were closer to the idea of science than were the services which I have been praised for.

Contrary to this Board, therefore, I ask myself whether we, the physicists, have not sometimes given too great, too indiscriminate loyalty to our governments, against our better judgment—in my case, not only in the matter of the hydrogen bomb. We have spent years of our lives in developing ever sweeter means of destruction, we have been doing the work of the military, and I feel it in my very bones that this was wrong. I shall request the Atomic Energy Commission to review the decision of the majority of this Board; but, no matter what the result of that review may be, I will never work on war projects again. We have been doing the work of the Devil, and now we must return to our real tasks. Rabi told me a few days ago that he wants to devote himself entirely to research again. We cannot do better than keep the world open in the few places which can still be kept open.

The hangings close.

The following text is projected on the hangings:

ON DECEMBER 2, 1963, PRESIDENT JOHNSON PRESENTED J. ROBERT OPPENHEIMER WITH THE ENRICO FERMI PRIZE FOR SERVICES RENDERED ON THE ATOMIC ENERGY PROGRAM IN CRUCIAL YEARS.

THE RECOMMENDATION FOR THE CONFERMENT WAS SUBMITTED BY EDWARD TELLER, THE PRIZE WINNER OF THE PREVIOUS YEAR.

Curtain.